For Miriam, Sture & Göran Rådström,
Compliments of Erland Anderson.
Om ett nytt liv i den Nya Världen,
Lars Nordh—, Beavercreek 27-05-99

Making It Home

Making It Home

Lars Nordström

Illustrations by M. J. Pfanschmidt

PRESCOTT STREET PRESS
PORTLAND, OREGON

Text copyright © 1997 by Lars Nordström.
Illustrations copyright © 1997 by M. J. Pfanschmidt.
First Edition.

A portion of this book was published in slightly different form in the anthology *The Prescott Street Reader*. The author wishes to thank Erland Anderson, Vi Gale, John Laursen, Cynthia Nordström, Martha Pfanschmidt, Mark Thalman, and George Venn for reading versions of this work in manuscript; their insights and suggestions helped make it a much better book.

Making It Home has been designed and produced by John Laursen. It was printed in the United States of America on acid-free paper. All rights reserved. No part of this book may be reprinted or reproduced in any form without written permission from the publisher:

> Prescott Street Press
> Post Office Box 40312
> Portland, Oregon 97240-0312

Library of Congress Cataloging-in-Publication Data
Nordström, Lars, 1954–
    Making it home / Lars Nordström; illustrations by M. J. Pfanschmidt.
        p. cm.
    ISBN 0-915986-27-2 (alk. paper)
    1. Vineyards—Oregon—Beavercreek. 2. Viticulture—Oregon—Beavercreek. 3. Wine and wine making—Oregon—Beavercreek 4. Country life—Oregon—Beavercreek. 5. Nordström, Lars, 1954– I. Title

SB387.76.O7N67   1997   97-1278
663'.2'0979541—dc21   CIP

Making It Home
is dedicated to the memory of my mother,
Hilma Sofia Nordström (1918-1986),
who insisted she would never visit the United States.

One

People ask me why I have come to settle in the United States. Some obviously want me to say that I think that this is the greatest nation in the world, perhaps because they really believe it, or perhaps because it would reaffirm something they are no longer so sure of. Those people are not interested in getting other answers, and it is easy, sometimes even embarrassing, to see the disappointment in their eyes when I don't tell them what they expect. Those who are critical of the United States wonder why I have left a place like Sweden to come to such a backwards place as this.

I tell them all that fate and love of a particular woman born on this continent conspired to bring me here. That is always a perfectly acceptable answer. I smile at the fact that everyone likes the idea of love being so powerful that it can take one's life beyond one's control, and even though it is true, shadows of a more complicated answer sometimes flicker in the back of my mind.

My wife Cynthia and I, and our two sons Nils and Carl, have lived on these eight acres saddling a small ridge for ten years now. We are still just getting acquainted. I am here to bring rush and movement to an apparent standstill, to allow for an inner settling of old, stirred-up debris. Days go by when I don't leave at all, just crisscross the land on foot in a steady string of projects, pushing materials in a wheelbarrow or carrying tools in buckets. Daily tasks within the wheel of seasons give the mind a certain elevated clarity. But then, sometimes, I just take a break from what I am doing and make an effort to stop my thoughts altogether, open myself up and invite the outside in. When you listen hard you can join another.

Of all the views here I like the one of the hills and the mountains the best. Many times a day my eyes wander across them: Highland Butte, Goat Mountain, Seosap Peak, Bracket Mountain, the entire Molalla River watershed. When it snows up there, the clearcuts stand out like large white rectangles, and the logging roads trace the contours of the folds like delicate, surgical incisions. The view of these different patches of the rejuvenating forest, in a good afternoon light, is a beautiful quilt of green and white hung on the wall of the world.

At midday I go outside to get a load of firewood from the pile behind the chicken coop. Loading the wood into a big canvas bag I notice unusually symmetrical cirrus clouds drifting in from the east. I think of Strindberg, who kept a daily cloud journal one summer in the Stockholm archipelago. He suspected that there was a correspondence between the shapes of the clouds and his mental states. At the end of each day he would carefully record his emotions and thoughts, and make detailed drawings of the dominant cloud type. But at the end of the summer his observations proved inconclusive and his mind moved elsewhere. Still, stoking the wood stove, I can't shake the notion that I felt just like those cirrus clouds as soon as I saw them.

It has been a cold winter. Nils and Carl found a frozen brush rabbit in the vineyard today, in the same area where Cynthia had seen one during the snowy weather six days ago. She kept pointing it out to me, "There, three rows down, just where those big weeds are. Don't you see him? He is sitting down, between those two plants!" I looked and looked but saw nothing, not having seen him move.

Today his body was still frozen solid, rigid in a sitting position, eyes open, as though ready to take off if frightened. The icy chunk of his body made me think of a story an aunt had told me, of a neighbor boy she had known when she was a girl. This was up in the north of Sweden, before cars. The young man had skied to the next village to visit a certain girl, and had stayed late talking and drinking too much. On the way home he had stopped to rest, and that was how they found him the next day, sitting on the trunk of a fallen tree, eyes open, face relaxed, body frozen all the way through. He had not even tried to make a fire. The parents had to thaw him out in order to straighten the body and get it into the coffin. Once in the coffin the body had been stored in the woodshed all winter, as was customary at that time, until spring made grave digging possible.

I take the frozen rabbit to the compost bins over by the garden, with Vidar, one of our cats, following. Later in the day he has eaten one entire back leg.

We first called our vineyard Beavercreek Vineyard because we live outside Beavercreek, and thought that the beaver was a fitting emblem in a state flying a golden beaver on the flag. Then I discovered that there already was a Beavercreek Vineyard in existence, which is not surprising considering the number of places in Oregon called Beavercreek. I called the owners of the original Beavercreek Vineyard, and they did not like the idea of us assuming the same name at all and elaborated on legal rights and attorneys. So we began tossing ideas around for a new name. For weeks we swung back and forth between the serious and the silly: Clos de Sapins. Vinland. Gopher Crest. Castor Creek. Cynthia & Laurentius. Skookum Hill. Ad Infinitum.

In a dictionary of literary terms I finally found a new name that sounded just right, Epyllion. An *epyllion* is a little epic which belongs to the genre of short picturesque poems. It is considered an idyll presenting an episode from the heroic past, but it stresses the pictorial and romantic rather than the heroic. Often it involves love between shepherd and shepherdess. Still, in spite of its definition, literary scholars argue that the term is really useless, so what could be better than that—a really useless term full of agreeable meaning.

Freezing rain starts falling in the evening, coating everything with a quarter-inch layer of clear ice. I never thought that a gravel driveway could become slippery, but it can. At night it looks like the uneven surface of exposed-aggregate concrete, polished and shiny. I have to walk in the grass in order to stay on my feet when I go down to close the gate for the evening.

When Cynthia and I started planting our vineyard, we planted one half to the peasant French-American hybrid called Maréchal

Foch and the other to the elegant pinot noir. Maréchal Foch, named for a famous French general of World War I, has grown beautifully, overcoming all obstacles created by two novice farmers. But the finicky aristocrat from Burgundy has given us nothing but trouble. It started when I bought the plants, sold as excess nursery stock from one of the major vineyards and wineries in the state. Many of them had cracks and very poor roots, and half never even sprouted in the spring. Of those which did, only a handful did well. In spite of promises of replacement plants I received nothing, sensing the shark teeth behind some of the wine industry smiles. Ever since then we have nursed the veteran survivors along, replacing the failures, fighting drought, weeds, gophers, and chilly temperatures. Ten years later, still without a significant crop, we conceded defeat. Three quarters of the pinot noir has already been dug out, with most of the land going back into pasture for a few years while we reconsider our options.

Two years ago we planted half an acre to a new variety called St. Croix, developed by grape breeder Elmer Swenson in Wisconsin, and another half acre to a dozen different varieties. They are all hybrids that survive almost arctic temperatures, and like Maréchal Foch, they are resistant to phylloxera and powdery mildew, which will eliminate the necessity of spraying sulphur as a fungicide all through the spring and summer. Eventually, when we will have had a crop for a couple of years and seen what kind of wine these varieties can make, we will select the best and fill in the space where we started with the pinot noir.

Having to change the vineyard from our original vision is part of discovering and accepting the reality of this land. We are located high, almost eight hundred feet above sea level, and in winter most of the vineyard land is directly exposed to icy air masses from the interior of Canada which come down east of the Cascades and funnel out of the Columbia River Gorge. Changing our ideas is also the result of our growing understanding of what organic, sustainable viticulture ought to be. I have a suspicion that farming must be a kind of lifelong listening to what the land says.

The full moon comes majestic over the frozen Cascades. I am down with a cold and cannot sleep in the blue winter light. I toss and turn and think about what I have done, quitting the job that I have had for more than four years to take a sabbatical. I was a technical translator and in some ways it was a good job. It came to me not of my own volition, as an unforeseen phone call with terms too good to refuse. It took years to realize it did not take me where I wanted to go, even though it made this perilous freedom I have chosen possible. During that time, as I commuted away to work, something I thought existed here became more and more difficult to experience. I trust it will speak to me again.

Thoughts meander to the familiar resting places, making well-traveled paths. Hours pass. The winter moon slowly drifts across the open spaces; the house ticks and cracks as the wood contracts in the dropping temperature outside.

It is curious, but I notice that in the years we have been here, the temperature has usually dropped right around the full January moon. We saw that same pattern in Sweden too. And once in winter, during a partial eclipse of the full moon, the temperature actually rose about ten degrees centigrade. When the shadow on the moon had disappeared, the temperature fell back down again.

Early in the morning Cynthia goes to check if the water in the chicken coop has frozen, and finds the entire flock massacred. Bodies are scattered about, with blood sprayed on the walls and feathers covering the floor. Returning after dawn Cynthia and I count and find two bodies missing, the chicken wire pushed away from the ground, a dog scat in the yard, and white feathers everywhere. What a depressing silence ten dead chickens make. I load the strangely flat bodies into a wheelbarrow, throw a shovel on top and wheel them far into the vineyard where I dig a mass grave for them. All day my mood swings back and forth between fatalistic acceptance and anger. Later I drive in to Oregon City to rent a live-animal trap. In the evening I call my dog-owning neighbors to tell them what has happened, notifying them that I have a trap set.

The following morning I find Elinor, our other cat, in the trap. She has been caught, no doubt, by her own curiosity and the smell from the pierced can of cat food which I have used for bait. I let her out and set the trap again. Later in the day, while cleaning out the coop, on the top side of the small door leading into it, Cynthia finds hairs stuck, reddish in color, long, smooth and wavy, suspiciously similar to that of our neighbor's dog. I call this neighbor back. "They look like the hairs of your dog," I tell him, "Why don't you come on up and take a look and see what you think." He agrees to come up the following day as soon as he is free. In the morning the clear sky is a pale yellowish white and the ground is rock hard. As soon as they wake up, the boys tumble out in their pajamas to check the trap and come back yelling "We caught him, we caught him!" Cynthia and I hurry out with tea cups in our hands and look at the embodiment of guilt itself. Indeed, it is the dog we suspected. I put a bowl of water inside the trap, call our neighbor again, and wait for him to show up.

The trap sits in the shade of the north side of the building and I notice that the night's frost remains there. But I am still angry enough to leave the shivering dog where he is. The day goes by and, it being a holiday, I am surprised that no one in that large family misses him. At dusk my neighbor finally shows up with a resigned look on his face, quickly agrees to reimburse me the five dollars a chicken I ask for, then leads his stiff dog home.

Cynthia is in the middle of pruning the old vineyard. Pruning requires a creative approach. Even though every vine has grown differently, the end result should be the same: good canes for the coming year or, if there is no available cane, a rejuvenation spur so one will grow the following year. She comes through with a pair of loppers and makes the primary cuts, severing the thick canes from the trunk, planning and directing next year's growth. Then she comes through a second time with a pair of pruning shears, to trim loose ends, remove smaller canes in tight spots, undo last year's plastic ties, and rip the woody tendrils off the trellis wires. This kind of cutting can be done in any weather. Later, in or after rain when the wood is supple, she will come through again, to bend the new canes and tie them to the wires.

I prune the replanted vines, which vary in age. The bushy growth of the one-year-old vines is cut back to two buds, and the vertical two-year-old vines are shaped into straight trunks if they are thick enough at the top. If not, they are cut down to two buds again and have to start over. During spring and early summer they will send out some horizontal growth along the wires, and perhaps set some fruit. Pruning is the dialog we keep with each other.

The cold weather suddenly eases for a single day. The icy wind has stopped blowing and there is no frost in the morning. An immense mass of warm air, the one called the Pineapple Express, has unexpectedly hurried in from some far southern place in the Pacific. It smells sweet for the first time this year and in the afternoon I open the study window for the first time too, trying to decide if the sweetness is the smell of the ocean or of the land it has traveled across.

One night on Crete, climbing a dark hill on the outskirts of Iraklion to visit the grave of Kazantzakis, Cynthia and I felt another wind blowing in from Africa, bringing with it incredible smells of the Egyptian desert. It was such a dry, spicy fragrance we could not force ourselves to return to the exhaust fumes of the ancient city below, even though we had an appointment. We stayed late, made love in the empty park, talking and watching the lights of the city.

The magic of sunlight warmed our house today. When we built it, we simply did what canny farmers in Sweden did for centuries: Found the right spot in the landscape and oriented the house according to the four directions, put most of the windows on the south side and practically none towards the north. No firewood, no attention, no ashes, no effort, no expense. *Tack sol!*

Going to get a basket of potatoes from the wine cellar, I look at the stone wall I built and notice how well the moss has grown on the rocks in just three years, without any help from a human hand. The stone wall faces north and trees above shade it. The stones around here are reddish when they come out of the ground, but exposed to the weather they start to darken.

Perhaps it is the chemistry of iron oxidizing. Today, the wet moss on them is a radiant green, in full bloom now, thin stems with tiny oval heads shooting out of the lush carpet. In ten years most of these stones will be completely black and covered with moss.

Potatoes—how suggestive and secretive they are in their bin, smooth and polished like glacial rock; how wonderfully they smell of the earth. I fill my basket with yellow Finns, the only variety we have left. The Nooksack is already gone and so is the Mandel, the blue German and Rote Erstling. We have to plant more. This is the best month of the year to boil yellow Finns, because their flavor is at a peak of richness, and they will not crack open, and it is too early for them to have started losing flavor and softening up in the process of sprouting. All winter they have matured towards a firm waxiness instead of that dry, crumbling texture they had in the early fall.

On my way back to the house, I think of dinner: boiled potatoes with dill and butter, some sausage, fresh bread, and a simple shredded cabbage salad with salt, lemon juice, and olive oil. Perhaps a Maréchal Foch of our first bottling to drink?

Tired of the six o'clock news I turn the TV off and step outside. It is already after dark and the evening has cleared into a cold stillness. East of our ridge a narrow tongue of fog crawls up the draw from the sea of fog below. The waning moon fills the fog with a strange fluorescent light. It will drop far below freezing tonight; I can already see the sparkle of crystals in the grass. I walk across the driveway to the lean-to shed where the thermometer is, and a neighbor's dog starts barking at the sound of my footsteps in the gravel.

After two days without precipitation I start removing the cut canes from the vineyard floor. Each row is about 220 feet long, with twenty-seven vines spaced eight feet apart. On seven acres there are 112 such rows, which would make a total of about three thousand vines if the whole vineyard was planted. Just to walk up and down all the rows is about four and a half miles. Now eighty-four rows are planted, but only sixty-eight with mature vines.

It takes several trips in each row to pick up all the canes and drop them at the end of the rows. Later I will load them into my tractor trailer. The work is monotonous and I get sore in strange places from all the bending down, but the body needs it no matter how the mind objects. In three hours I can clean about six or seven rows and haul the cuttings to the back where I have a huge pile of them. These cut vines are mainly carbon and water pulled out of air and soil; they are sunlight made heavy.

This is a job which must be done before the grass starts to grow and entangle the canes. For us, canes are really a resource. A few are cut into four-bud sticks and sold as cuttings for new plants, and every year women will come and gather long ones for wreaths. Others will get the thickest wood for the coming summer's barbecues, but most of it is chipped for use as mulch and chicken litter.

While pruning our fruit trees, I spot a solitary heron flying low, just over the rooftops, turning to the north and gliding over my neighbor's farm buildings before disappearing from view. My eyes follow his slow flight like a prayer to an old spirit.

Two

When I came of age, it was the United States that was the foreign focal point for my generation. The rest of Europe did not hold our interest. America was where our imaginations found their nourishment: we discovered it in music, film, and literature; in the pop art of New York; in the many dimensions of the counterculture and the anti-war movement, in hippies and yippies. 1968 had come and gone, and the U.S. had just lost the Vietnam War. What would happen next?

And then there was that other, more neutral dimension of the North American continent beckoning—a spectacular natural landscape beyond anything one could find in Scandinavia—the prairie, the Rocky Mountains, the deserts, and the forests of the Pacific Northwest, where uncle Hjalmar had been a logger. He had not made it rich, but he made it back with more money than when he left, and that counted for something. As a matter of fact, having gone there at all counted for something. He talked of black bears and fishing, bar fights and cars, and impressed us with sporadic expressions like "You big dumb Swede!" and "Holy smoke!" It was obvious that he relished his years in the Oregon woods, and more than once insinuated, after making sure that my aunt was out of earshot, that the biggest mistake of his life was the decision to return.

My aunt was my mother's sister, and my mother later explained that it was because of her that Hjalmar had returned home. But she added that he wasn't bitter, just inclined to reminisce about his big adventure.

Alone, after a day of bottling wine in our small winery with Cynthia and the boys, I carry twenty-odd empty five-gallon glass containers down into the cellar—one carboy in each hand, trip after trip. When I start the grass is still aggressively green, but, as the evening fades, fog slips through the damp, leafless trees. Night takes over, filling my arms with a dull, dark ache. Carrying the last carboy down everything is black.

The weather turns clear again with cold winds out of the east. At night it drops far below freezing several days in a row. After breakfast I load axes and wedges and a ten-pound sledgehammer into a wheelbarrow and get to work on a pile of cord wood. I give the wide slabs of a huge elm trunk another go, but they are hopeless. I manage to split a few where the grain is reasonably straight, but with all the others, the axe bounces right back from the twisting grain. I will have to cut those into chunks with the chain saw. It is the last time I will take elm, even if it is free and gives good heat. But the frozen maple and Douglas fir split like cedar and the work takes hardly any effort at all; I like the ring of frozen iron when a wedge is struck by a sledge.

It was my grandfather who taught me to split frozen wood. When our family visited my grandparents in the north for the Christmas holidays, I always wanted to split firewood and help stoke the furnace in the basement. Having grown up in an apartment with central heating, where the radiators were automatically adjusted for the seasons, one never thought about where the heat came from. There was not even a furnace for our apartment building; it was located in another block.

The floor of my grandfather's woodshed was covered with a thick layer of sawdust and bark, and even in the dead of winter it was full of the rich smells of newly split firewood. His axe heads were all sharp and firm on the handle, and he was always very particular about which axe to use. He showed me how to avoid the knots and aim for the cracks that showed in the ends; he taught me how to swing my arms in a long, powerful arch.

My grandfather always used to repeat that old farm saying: "Next winter's firewood should be cut, split, and stacked before spring planting," even though his oldest brother had inherited the family farm, and he and his younger brother were forced to leave for the city and look for work in the factories, and thus never had any spring planting to do.

Early in the month a small band of robins arrive from the valley below. I have seen the voluminous winter flocks expanding and contracting in the sky down there, the whirling of thousands upon thousands of robin and starling wings. This is nothing in comparison, only a handful of birds searching the vineyard for food. Some will like it well enough to nest here, and I know we will have to fight them and their offspring in late summer after the grapes have finished turning. But in the slanted afternoon light I hear them singing for the first time of the year, a few, clear melodious tones lifting my winter spirit out of my heavy, muddy rubber boots.

In Swedish my surname means "northern stream," and my first name is a shortened version of the Italian "Lorenzo," from the Latin "Laurentius," which came to Scandinavia with the Renaissance. In many ways mine is a good name to have in the United States, because people remember it easily. But there is another aspect to having a foreign name here, and it is the unending experience of finding it altered, as if the culture is permanently incapable of learning its correct form. I spell my surname with a letter that does not even exist in the English alphabet, the vowel "ö" in Nordström. Since few Americans know how to make the sound represented by "ö," they make the best of it and substitute "o" for "ö," which changes the pronunciation as well as the spelling. Similarly, "Lars" is not pronounced the way people think it is, but trying to explain it is too complicated and ultimately meaningless, so one leaves it, and is altered by the decision. Somehow it symbolizes the immigrant situation: just as one cannot bring one's name across intact, one cannot transfer one's complete identity either. Some part of the self, properly pronounced, is forever residing in that other world.

And this creates strange allegiances. The only TV reporter whose name I can remember is Lars Larson, and once I bought a book because I was intrigued by the fact that the author's first name was Lars—perhaps I thought I would discover something about my own situation. Then there was Lars Ahlström, whose cabin, built in 1902 on the western tip of the Olympic Peninsula in Washington, was the homestead farthest west in the continental United States. It is reassuring to know that this man moved closer to the edge of the continent than any other non-Indian immigrant, and then managed to hang on there.

In Sweden I do not think I ever saw my name misspelled; here I am perpetually amazed by the seemingly endless possibilities that my name provides. As an undergraduate, I was called Lars Wordstrom by one of my professors; once he even wrote Lars Wordstorm, and I never learned whether he meant that as an academic, etymological joke. Perhaps he thought that I talked too much and this was his way of pointing it out. One magazine I get is addressed to Lars Nordstorm, and I have received things addressed to Ms. Lars Nordstrom. Over the phone I have become Larz, Larce, Lar, even Laro; Norstrom, Nordstron, Nordstrum, Nordstum, and once even Mordstrom ("mord" means "murder" in Swedish). The Swedish Information Service in New York surprises me by keeping an alias in their files called L. Nordstöm, a mistake no Swede would ever make. I have never bothered to correct any of these variants, not even when a local bookstore put me on their mailing list as Lard Nordstrom, and I seriously started wondering if Americans ever stop to consider what constitutes a name.

I recently got an eight-inch seedling from a walnut off a tree northeast of Stamford, in Lincolnshire, England. It comes from an old tree, I was told, that has grown well, adapted to the cool weather and late frosts, yielding good harvests of medium-sized nuts. Today, in the small field behind our house, I planted that seedling.

If I am lucky I'll be fifty when I get my first harvest of English walnuts. The Chinese say there are three things a man must do in his life: have a child, write a book, and plant a tree.

The first surviving map ever drawn of our area, which is formally known as Township 3 South, Range 2 East, Willamette Meridian, was "examined and approved" by the Surveyor General's Office in Oregon City on August 12th, 1852. It shows a sparsely populated landscape. There are only a few scattered homesteads along the main road connecting Oregon City to Mulino in the south. That road is still there today. Beyond the road the forest appears intact. There are several Indian trails, but there are no signs of settlements along them. All the creeks are carefully drawn, as are some of the hills and steep ravines. Across the map, the surveyor Joseph Hunt has written: "Land considerably rolling. Soil 1st and good 2nd rate. Timber Fir, Cedar, Alder, Maple. Undergrowth Vine Maple, Hazel, Fern."

In his field notes on our particular section, he repeated what the map says in general, but noted that the original forest had been thinned out by fire, and he classifies the timber as second rate. Perhaps that explains that layer of charred bark and pieces of charcoal I sometimes find a foot or two down in the ground. The surveyor also notes an Indian trail going in an east-westerly direction across our section, disappearing up into the hills, and it does seem to correspond to stretches of three local roads here today. It is quite clear that no one farmed this land in 1852.

Towards the middle of the month the green of the ground begins to assert itself in earnest and push back the dull hues of winter. The white marsh hawk has returned again this year and flies by almost daily. The days are already noticeably warmer and the grass responds. I hear the first frog sounds from the pond across the road, wet-skinned amphibians croaking for love. This year it hadn't rained for a few days when the first sounds came, which surprised me. Perhaps they were asking for rain.

"By the way," the man from the phone company says, sweeping his screwdriver over the vineyard in a gesture of skepticism, "why are you growing grapes here—was it something your family did back in Sweden?"

"No," I say, deciding not to get into the climate and geography of Europe. "We had to farm to be able to build a house and live here. Because this is EFU-20, exclusive farm use, twenty-acre minimum, and this is an eight-acre parcel, we had to practice what the county officials call intensive agriculture. Things like nursery stock, flowers, berries, or grapes. So we decided to grow grapes. We have always enjoyed drinking wine."

"Is there any money in it?"

"We get a thousand dollars a ton for organic grapes," I say.

"That's good money! An uncle of mine grows pears up in Hood River, and pears are only five hundred a ton."

"It may be a good price per ton, but last year we harvested less than two tons per acre. As the vines mature, we might get more. With three and a half acres in production, it is not a great way to get rich quick."

This time he takes a while to respond. Perhaps he thinks that anyone growing grapes for that amount of money must be crazy, or maybe he does not approve of wine. He would not be the first one.

"I sure like the view you've got," he finally says, closing his tool box and the back of the van. "Anyway, I've fixed the wires that had come loose, and that should do it. Call us if you have any more problems with the phone."

I watch him disappear down the driveway. It is true, we did not have to buy this land, nobody forced us. Perhaps we did not really know what we were getting into, but we did choose it.

The afternoon is damp after rain but the air feels relatively warm. Across the road fir trees comb and card lingering remnants of cloud as if it were wool. For the first time this year, it is comfortable to stand quietly outside as evening falls, watching thin insects in the air rise and fall in an endless circle. What is the purpose of this up-and-down dance? It is more beautiful than ballet and their musical hum is far more subtle. This vineyard is the home of these insects. Maybe an entomologist has a name for them, but they have never needed it. They know who they are. This evening is their life. The president's message to the nation is irrelevant to them. Behind their dance, watching the darkness overtake the Cascades, I imagine all the sheets of rain that fall on the mountains, the water seeping down from up above, moving minerals and trace elements, passing through this soil where the vines eventually search it out and load it into the grapes.

I remember a floor mosaic in the ruin of a Roman villa I once saw on Cyprus. There, a seated Dionysus, crowned with a wreath of grape leaves and holding clusters of grapes, has handed the reclining nymph Akme a bowl of wine. There is no doubt in my mind that the ancients knew that the gift was sacred. Living here I am beginning to understand that I have always wanted to touch the flow of life symbolized by that image, that what I thought were mundane things like food, shelter, and daily routines actually belong to a sacred river that flows below us. When you start eating and drinking out of your own hands, you gradually discover that you are flowing through something much larger than yourself, that something is holding you, carrying you as you hold it.

A late wet snow surprises everyone, we who thought winter was over. Almost all the robins and Oregon juncos disappear overnight. There is deep snow everywhere in a fifty-mile radius, so where do they go? Nils and Carl put bird seed in the feeder but not many birds show up. All work ceases outside and Cynthia cannot prune the last of the grapes. The boys go sledding down the hill behind the house and I put on rubber boots and head towards the high point of the vineyard to view the valley. A deer has come through an opening in the fence and meandered through the lower section of the vineyard. I wonder if this is a normal visit, which I would never have known about if it had not been for the snow, or if it was one driven by the sudden scarcity of food. Whatever reason, the tracks speak their own story. I follow them up the hill and see how they turn and return back out through the same opening in the fence. It was an unfortunate place to look for food, since the snow is deeper here than in any other place in the vineyard, the storm having come in from the southwest and piled it up against this slope.

Standing on the high point, seeing a white valley all the way over to the coast mountains in the west, I get a sudden sense of an unfathomable power. Nature could so easily shut down our small human enterprise. I shiver and feel transparent, as if the icy wind hunting over the snowy ground blows right through me on an errand where humans have absolutely no say.

We have a few small sections of the vineyard that are fenced in, and patrolled by two flocks of geese. It is an ongoing experiment to see to what extent they will be able to become our weed control. The wet springs in Oregon create a problem for the grape grower: abundant growth on the vineyard floor. One can mow it all the time, disc it, or kill it with herbicides, but all those methods have their drawbacks. Mowing means that one of us has to spend long hours on a tractor. If the ground is kept tilled, it will erode the topsoil. Since we are organic growers we do not use herbicides, and with them, the erosion is even worse. Foot deep gullies in the surrounding Christmas tree farms are not uncommon, and all winter the ditches around here are filled with chocolate-brown runoff.

Or, one can look at this abundant growth as food. In Beavercreek, geese can live on a solar-powered diet of grass, weeds, and water nine months out of twelve. They just need to be fenced in, and are hardy and healthy and a great deal of fun to watch. Unlike tractors, they reproduce free of charge. They are edible. And more important, they improve both soil fertility and soil structure through their droppings, which benefit worms, whose tiny tunnels counteract soil compaction and improve the ability of the soil to absorb rain water.

The geese look surprised at the late snow too, which covered up the scant grass, and they wander around hesitantly and aimlessly. I spread some straw for them to rest on, and pour some grain in a metal trough. They hiss and honk as I work among them, and I can see their childlike breaths forming in the cold air. Later, I notice that they ignore the straw and nap right on the snow. I marvel at the ability of their naked feet to walk on ice crystals.

My father laughs at the lifestyle Cynthia and I have chosen, but it is a kind and gentle laugh. He says we remind him of the characters in Moberg's epic of Swedish peasant emigrants, pioneers who came to transform the unbroken North American continent into farm country before the turn of the century.

"I understand why people did that a hundred years ago," he says with real concern in his voice, "they did it because they were poor, starving, and landless. But today? Aren't you a century too late? I don't know what most immigrants do today, but I bet they're not becoming farmers. You have an education. And have you really thought about the boys, how it will prepare them for the future? Isn't your life just a form of escapism from modern society?"

It is not the first time I have this conversation with him, and I let him talk. We both know that it can't be resolved over the phone; the subject is too big. He sees the future his way, and I understand why; for him, his career and success was tied up with his own migration from farm to city, and he sees the future in those terms too. To him, Cynthia and I have chosen a life that suggests the past. But he is not talking to make me change my mind, just to rouse me to a reaction. He knows about my computer and fax machine, and he knows that because of them, I am able to work at home.

"I'm not sure I like everything in modern society," I answer, "because if we look at it in terms of sustainability, I'm convinced it is headed in the wrong direction. And just because most people go along with it, doesn't make it the right thing to do. A small farm, on the other hand, makes you very aware of the limits of things, of food, resources, energy, the environment, and those are the issues of the future. So when it comes to challenges, I think the opposite of what you said is true."

"The opposite of what? I'm not sure I understand," he says.

"That a small farm is a good place to prepare the boys for the future. I think the tradition of self-reliance inherent in farm life has much to teach them.

"Besides, farming and translation is a great combination. If it's sunny I can work outside, and if it is rainy I can work at my desk. It provides balance."

"Well," he says, "it is your life. I was just thinking about the boys."

"Yes, and so am I. But I'm afraid that we'll have to resolve this later."

"Yes, I suppose so," he says.

"Take care!"

"*Adjö med dig!*"

Every year we grow a little bit more of our own food; every year it nurtures us a little further. Tonight's ragout is entirely our own, with Blue Lake pole beans, corn and peas from the freezer, home canned tomato sauce, handfuls of shallots, and a head of garlic from the basement larder. The lamb ate the grass in the vineyard. The wine flavoring the stew is from a bottle of our '91 Maréchal Foch, our least appealing vintage so far and now relegated to cooking wine. Only the wheat in Cynthia's bread is not ours; on the sack it says Brigham, Utah. I have never been to Utah, but Brigham has often come to me.

Working in the kitchen, I think of the life of Angelo Pellegrini, who came to the Pacific Northwest as a young boy from the poverty of the Tuscan countryside just before World War I. I too share his lifelong feeling of awe observing the natural abundance of this region, and the surprise and disbelief he felt whenever he saw a blindness to it. Every book I have read by him is such an eloquent attempt to shatter that blindness and inspire the enjoyment of the earthly delights of the bountiful garden. Beneath his academic disguise he will always remain the wise old peasant, proponent of the three-fold imperative: "Grow your own vegetables, bake your own bread, make your own wine." His advocacy of a balance between the cultivation of the garden and the mind, the palate and the spirit, the earth and the soul, has affirmed so much of my own thinking. The virtues he espoused—frugality, work with one's hands, humility towards one's fellow human beings, careful husbandry of worldly possessions—I embrace them all.

Angelo, many years ago, after a great meal at my neighbor's house that came out of their garden too, we almost called you. My host got as far as getting your number in Seattle from directory assistance, but then, considering the lateness of the hour and the amount of wine consumed, we decided to put it off. Neither of us ever called you. Now you are beyond calling, and even though you and I never spoke, I know you will accompany me for a long time.

Spring is signalling everywhere. The garden, which just a week or two ago was dull and faded like colorful clothing washed too many times, is now thoroughly green with weeds. The garlic Cynthia planted in the fall is pushing vigorous green spears, and there is a small flash of blue rosemary petals—first to flower in the garden—to which my eyes keep returning. Snowdrops and crocus are already in bloom, as are the maples, and buds are swelling on the almond trees, lilacs, and currants. The vines look as if they are still dormant, but all the fresh pruning cuts bleed now, revealing intense hidden activity.

Three

I went straight through school until I was nineteen, and a few weeks after my graduation I was drafted. It was early summer when I was released after eleven months of military service, and life seemed full of possibilities. I had no obligations and no definite plans for the future. I was too restless to go back to school, and I wasn't sure about what to study anyway. Traveling was another option. And then there was the idea of moving to the country to start a communal farm; the back-to-the-land movement was in full swing and its vision of a slower and more balanced life had swept me with it. If enough people moved back to the country perhaps everything would change. Perhaps we could change society for the better. Some of my friends had been looking hard for a place in one part of the country where there were several communes already, but nothing had turned up. While mulling these things over, I found a job as a desk clerk working the graveyard shift in an old Stockholm hotel. Usually, there was not a whole lot to do, and I could spend the nights reading, napping, or talking to jet-lagged Americans who could not sleep in the light summer nights.

After a month the regular night desk clerk returned from his vacation, and I got hired on for the swing shift. There was more work to do, but the pay was better and by now I was familiar with the routines. Less than a month into my new schedule, a very attractive woman my own age walked through the doors and asked me if there was a vacant room. I answered yes. She asked if she could see the room, and I answered yes again. Finding the room OK, she checked in and disappeared to get her luggage. Later in the evening she stopped by the reception desk to ask me how to get to an American movie. We never stopped talking. When my shift ended at ten o'clock I asked her out for a beer. Without knowing it, within a few hours, both of our lives had changed direction.

Cynthia peruses the catalogs from various chicken hatcheries trying to decide what to get for the coming year. What do I think? Should we get new hybrid breeds or old varieties? We pore over pictures and descriptions, an entire chicken universe of meat birds and layers, bantams and heavy breeds, eggs of different sizes and colors. It is hard to choose, but we finally settle on buff-colored Orpingtons from a local hatchery for our eggs, "a quiet English fowl," the catalog says, "an old breed not bothered by cold weather, good layers of brown eggs." For meat we decide on something called a Cornish cross, a recent hybrid of fast growers that don't like to exercise and don't watch their weight.

Inspired, I start working on the vacant coop. I dig a fairly deep trench along the perimeter of the roofed area, then frame a narrow concrete form inside it. Once the concrete footing has been poured, no predator will ever be able to dig its way under again. Later I bolt a wooden sill to the top of the cement footing and staple the chicken wire walls to that. Once it is done, it looks impenetrable. When a friend sees it completed, he dubs it "Fort Chicken."

Days of mild east winds from the desert interior dry the raised garden beds, and Cynthia begins weeding volunteer lettuce and bread-seed poppies, chickweed and dandelion, grapes sprouted from seeds. Soon she has a great stack of fresh weeds and it is time for me to attend the compost piles again. All winter they have been left to their own slow burning under the tarpaulin; nothing has been turned and only the kitchen scraps have been added. I have five compost bins in a row, and basically add new material in the bin at one end, and when it is full, throw the entire pile to the next. Ten or twelve weeks later, after having been turned and mixed five times, the compost is ready at the other end.

To make room for the new weeds I start sifting the old, finished compost through a wooden box frame with half-inch hardware cloth. Whatever passes through the mesh into the wheelbarrow gets spread out in the garden. The pieces too large to pass through the mesh—corn cobs, twigs, decaying bones, the woody stems of cabbage, broccoli and sunflower, peach pits, slowly disintegrating wine corks, mussel and oyster shells, dirt clods, locks of hair and fur, chicken and goose craniums, root balls from unknown plants, halves of avocado seeds—are returned to the first bin for another journey through the system. The boys are fascinated with the things that get caught and come over every now and then to study the contents. Nils retrieves a rooster leg bone with a spur on it for his treasure box. It takes almost one day to sift and spread ten wheelbarrow loads of dark brown compost. The following day I throw each of the old piles one bin over, eventually creating an opening at the end.

New piles are layers of anything organic that comes along: fresh weeds, the large chunks of old compost, chicken manure, wood ashes, handfuls of ground limestone. I stack the layers around a wooden post, and sometimes I add a touch of water to get the humidity right. It is topped with a coat of chipped grape vines or straw to keep any surface seeds from sprouting. The post is then removed so that air can reach the inside. In a few days the temperature inside the pile will reach 160 degrees Fahrenheit and steam will rise through the vent. Nils and Carl fill an empty magnum bottle with water, cork it and tie a string around the neck before they lower it into the belly of the pile. Later they pull it out and laugh as they wash their hands with "free hot water."

It is surprising what microbes have dissolved in our pile: the cotton filling of an old futon mattress, a broken straw hat, a worn out seaweed fiber door mat, discarded wool, old burlap sacks, broken wickerwork baskets, cotton rags, and worn-out work gloves. Everything is humbly transformed into the magic substance that powers the garden; compost is the essence of organic gardening, creating a healthy, biologically active soil rich in humus. The word humility, someone said, derives from the word humus, and here, where spent life is transformed into new, one discovers it again and again.

Topping off the barrels in the cellar I am pleased by the changes the six-month-old wine has gone through. This time of year it seems to cross some invisible threshold and become smoother and softer. It feels richer in the mouth. In the new barrel I can, for the first time, taste a mild vanilla flavor. The sharp edge that used to alert the tongue is almost entirely gone. Perhaps it is the result of the malo-lactic bacteria converting the sharp malic acid into the more mellow lactic acid, even though the cellar has been rather cool for that kind of activity. Maybe some of the acids have just finally precipitated out. Whatever the reason, young Bacchus straddling the barrel has just raised his beaker.

The years between the first map of 1852 and the first plat map, drawn in 1860, show a great influx of settlers. Every single land claim in the township, except one, is by a man with an English name. The exception is a French name, perhaps a retired trapper from the Hudson's Bay Company. In spite of the influx of people, about a third of the plat remains open public land. No claims have been entered for our section, and it lingers white and empty.

In the next plat, of an unknown date, claims have been registered in our section. The place where we now live was first granted to a couple, Timothy and Martha Gard. I do not know if they ever lived anywhere on their three-hundred-acre parcel, and I do not know if they were the ones who had it logged, since all of Beavercreek was logged in the 1880s. Or the logger might have been John K. Landeck, to whom the land passed in 1884. From him it went to Louise Kamrath in 1901. The Kamraths were the ones who finished clearing it after the logging. The house across the road is known as the old Kamrath place, and the road coming up to our ridge from Beavercreek still carries their name. The land stayed in the Kamrath family until 1976, when it passed on to the man we bought it from ten years later.

The first trees that Cynthia and I planted on our property, two Hall's hardy almond, also turn out to be the first to bloom each year. They start slowly, with a few new flowers every day until the mild temperatures push the trees into soft, pink explosions. The trees have bloomed for three seasons, but only profusely this year.

Late in the afternoon, I walk down to the trees to watch the insects work. There are only a few regular honey bees, and some that look similar to flies, but they are skinnier and move differently. Maybe they are the local Mason bee. The whole tree is electric with activity. It is amazing how much insect fuel a small tree like this puts out during bloom, how much life it energizes. As I sit watching in the fragrant air, the first hummingbird of the year buzzes in to feed on the flowers. I try to determine what kind it is, but move carelessly and scare it off before finding out.

This spring we are planting yet another selection of French-American hybrid grapes in our vineyard, all relatively unknown varieties. In our nursery we have grown four whites, Seyvál blanc, Vidal blanc, Orion, and Totmur, and two reds, Oberlin noir, and de Chaunac. They are all supposed to ripen early, be resistant to phylloxera and powdery mildew, and be winter hardy. But will we like the wine?

The buds are swelling on the grapes in the nursery, and planting is getting urgent. After digging the vines and carefully labelling the different varieties, Cynthia dips them in a bucket of water and brings them down to the vineyard. As soon as I am done digging the nursery, I join her in the field. The hours extend into early evening as we work along each planting line. Just as we are about to finish, a purple balloon comes dancing a few feet off the ground like a big, happy, succulent grape. We are too tired to even attempt to catch it for the boys, and simply watch it sail through the newly planted vineyard and whirl a few times at the far fence before climbing back into the sky and disappearing southward.

One of the mother geese has arranged a brooding place inside a small plywood nesting box. Every now and then she goes inside and pokes around in the litter preparing to lay an egg, while two of the ganders guard outside. Occasionally I catch her sitting in there by herself, as if trying it out. The ganders have become extremely noisy now as the number of eggs increase.

One day when the whole flock is feeding in a different part of the enclosure, I jump the fence to take a peek. The nest is covered with a carefully arranged layer of shredded grape vines with some feathers and down mixed in. Konrad Lorenz discovered that geese covered the eggs not to keep them warm, as was previously thought, but to hide them from egg-eating crows. I gently grope around and find four eggs hidden underneath. The eggs are huge in comparison with chicken eggs, heavy and easily twice the size. I cover them back up and get out from the pasture. Later Carl checks the nest again, and the number of eggs has increased. It works out to about one new egg every other day.

Soon the grasses and the weeds will burst, and if not kept in check, will easily grow six feet tall by late June. So this time of year we have begun to combine our weeder geese with sheep, for the simple reason that the geese cannot keep up with the growth. The geese have also proven to be picky eaters, and sheep seem to like what geese leave behind, like English plantain and oxeye daisy. We are experimenting with an old, short-legged English breed called Southdowns. Ideally they should not be able to reach the grape leaves growing above them, but be forced to eat only what grows on the ground. Miniature sheep might be ideal, especially if the trellis is modified to raise the canopy.

Last year we had two wethers, this year five rams. As soon as I open the back of the pickup and let them loose in the vineyard they start eating. They eat in the rain. They eat when it is windy. They even eat in moonlight. At the end of the season we eat them.

Living on a vineyard, I like the French paradox, which says that the French have a surprisingly low rate of heart problems in spite of their high fat intake. Several studies point to the consumption of red wine as the explanation for this, while others say it is simply due to the fact that the French also eat such voluminous quantities of grains, vegetables, and fruits. To be safe, I have decided to do both, though vague feelings of guilt sometimes manage to rise from that old Lutheran suspicion of anything related to pleasure, a trait I seem to have inherited from my formative years. "What?" the voice says, "a glass of wine for dinner every day? Sinful! You might become an alcoholic!"

But I am learning how to answer this voice. I patiently explain that food and wine belong together, that wine in moderation can be good for the body. Not only is there a pleasure in savoring a dinner where the different courses are brought to the table piece by piece and the conversation is allowed to meander, but there is also satisfaction in drinking what we have grown and made ourselves. Wine is a living thing that not only tells you of what it has gone through since you last tasted it, and how it compares with other wines made from the same grape, but it also reminds you of something important about life: it has to be celebrated!

The rain finally lets up for a day and I go for a walk in the woods. It is damp under the dripping trees and the forest floor is soft. In the rich smell of moss and ferns, fungi and decomposing wood, I think of the Molallas, whose obscure story I have been reading about, scattered fragments pieced together by anthropologists Zenk and Rigsby. Our children at Beavercreek Elementary learn about the pilgrims at Plymouth Rock, pioneer life along the Oregon Trail, details of the Civil War, but they have barely heard of those local tribes whose history stretches back for thousands of years, and who subsisted here only a hundred and fifty years ago.

As far as I understand from the early maps of the Northwest Indian tribes, we live in Molalla territory. Once, the tribe ranged along the entire western and eastern slopes of the Cascades, from the Klamath basin to Mt. Hood. Up in the mountains, they maintained a network of trails, and lower down, used dugout canoes to travel the waterways to visit and trade. They were known as great hunters of deer and elk, expert users of dogs to track and drive the game. They wintered on the west side of the mountains. A northern band used to camp at Dickey Prairie, a place not far south of here. It is a beautiful, bowl-like area with open fields and oak groves, surrounded by hills through which the Molalla River flows.

Like their Chinook and Klamath neighbors, they would press and flatten the foreheads of freeborn girls to exhibit their social standing and make them attractive wedding prospects to these tribes. The Molallas raided the Kalapuyans for slaves, and were in turn raided by the Cayuse and the Nez Percé. They cremated their dead.

When the white settlers arrived, their clearing and farming rapidly transformed the entire ecology of the Willamette Valley. The camas meadows vanished under the plow, and the salmon and steelhead runs started their long, gradual decline. Birds ate the huckleberries and the tarweed seeds. Already by 1900, half of all the native plants in the valley had been replaced by foreign ones.

By then, most of the Molallas were gone too. Malaria, smallpox, measles, and several other diseases had preceded white settlement and taken a heavy toll. Following the Cayuse killing of the Whitman missionaries, there were reports of apprehensive settlers having massacred a large number of Molallas at the battle of the Abiqua in 1848. The Champoeg treaty of 1851 was never ratified and Molalla claims were ignored. The ratified treaty of 1855 provided for the removal of the Molallas to the Grand Ronde Reservation. In 1915, their chief, Henry Yelkes mysteriously died—or was murdered—in a local celebration, symbolizing an end to the Molalla presence.

Returning through the woods, I know that our vineyard is an inextricable part of this continuing process of altering the original ecosystem into something very specifically European. Yet I see myself as a caretaker here, tending the land for the future, as best as I can, in the absence of the rightful owners, whose culture was pushed into extinction long before I arrived.

Three days before the vernal equinox, the wheel of time completes another revolution and I am a year older. I have always thought that the date corresponds to the composition of my psyche, with light and dark moving toward balance, but with the scale always tipping in favor of darkness. I have now been married for as long as I have lived unmarried, and half of my married life I have been a father. I have lived longer with Cynthia than with my mother. It feels like the equinox of my life.

In the evening I am treated to a farm feast: spring salad greens from the cold frame, freshly baked bread, a roasted leg of lamb pierced and stuffed with innumerable cloves of garlic and coated with a paste of cumin, garlic, and olive oil, baked potatoes, and apple pie from our canned apple slices. I pour a bottle of our first vintage of Maréchal Foch, harvested with such gratefulness after four years of vineyard labor without a crop. As I drink it, I think of the yellow grass and the dry, red harvest soil, Cynthia's sleeveless white shirt and her straw hat, her tanned arms among the green leaves, and her stained hands reaching for the small blue clusters. I even remember the smell of her skin in the hot sunlight.

A friend in the Eola Hills, located south of us, mid-valley, reported violet green swallows two weeks ago. But we live on the fringes of the great mountain forests, and I suspect that there is little food for swallows there or in the surrounding Christmas tree farms. Perhaps that is why they are late.

Today a handful of them finally appeared. I am amazed that they can find their way across all those miles from Mexico and Central America. How can their small brains remember such a large map? Ornithologists have shown that some swallows return to the exact place of their birth. I wonder if the ones that came today are the ones raised in our bird houses last year, but a few days later they are gone again, returning to that place on the map within. I let my eyes sweep over the vineyard and the surrounding hills, thinking that I too seem to remember this from somewhere.

Quitting my job involved removing myself from the translator lists kept at the numerous agencies I had worked for, but several of them keep calling me to do small jobs for them. Obviously I have not been removed from their lists. It is flattering, and I get to practice saying no. I tell them that I am busy with other projects, and won't take on new work.

"Oh, but this is such a small job, it won't take you long at all," they try. "You have several days to do it. It's ready to go out the door this very second."

"No," I remain firm. "I want to finish what I have started first. When I have the time and want to take on more, I promise I will get in touch." Most give up at this point, and ask me instead if I could recommend another English-Swedish technical translator. A few ask when my project will be done and when I will start accepting work again.

"Well, I don't know when it will be done," I answer vaguely, "it seems kind of open-ended at this point." Since almost all technical translation work involves new products and much secrecy, there is an unwritten rule not to ask too many questions.

"Oh," they say, "open-ended, I see. Well, good luck to you!"

"Thanks!" I say and hang up, wondering what they would say if they knew that my project involved translating a dead American poet without even knowing if anyone will ever publish or pay for my work.

Towards the end of the month I have a vivid dream about deer returning from the lower elevations. In the dream I am sitting in a high place, hidden, watching a solitary animal wander through the vineyard nibbling both the vines and the weeds on the ground. It is a doe, but I think that there is a fawn somewhere too, maybe other deer as well, and I keep still in my hiding place. I sense that the doe is aware of my presence. Maybe she has smelled me on the wind. She is not frightened, having come to survey food and shelter for the coming season.

Even though I realize that the vines have not leafed out yet, and that my dream was of the future, I still tiptoe around all the windows on the second floor with a view of the vineyard, looking carefully along the edge of the woods. No deer. Almost every day for more than a week, dawn and dusk, I go through the same routine at the windows.

Four

Cynthia soon returned home to California, and I suddenly had someone I would very much like to visit. A few days after dropping her off at the airport I got a call from my friends: they had found a farm that had just been advertised in a local paper. It was perfect; the price was reasonable, and immediate action had to be taken to secure it. I managed to arrange a few days off from the hotel and the next day I was on the train. It was a beautiful farm right on the Norwegian border, located on a small rise above a lake, with pastures and fields sloping to the south and to the west. It looked like something out of a children's book. The old farmhouse was large and in good condition, and there were several barns and outbuildings. Returning to the city, I didn't know where to go—to California or back to the land?

In the end, I managed to choose both. I had inherited some money from my grandmother, and it was just sitting in a bank account waiting for something meaningful. I could put that money into the farm, and keep working at the hotel until I had enough for a trip to California. So I decided to buy a share in the commune first. The farm was quickly purchased and my friends started making plans for moving there. I wrote Cynthia and told her what had happened, but that I would still come to visit her as soon as I had saved up enough at the hotel. At the time the U.S. dollar was a weak currency, and the exchange rate with the Swedish krona was in my favor. By the middle of October I was ready to leave.

In the evening, after two days of heavy rain, there is an absolute saturation of moisture in the still air, a rich smell of moss and earth. I am reminded that this used to be a temperate rain forest. As I stand there breathing, it feels as if I were coating the inside of my lungs with vapor. It is not an unpleasant feeling. Deep in the great pile of pruned vines beyond the pasture, two frogs warm up for another nocturnal symphony. I listen for a while as the dampness permeates my clothes. There are roots growing in the darkness.

The mother goose in the plywood nesting box started sitting on her eggs two days before our small earthquake, then got off for a few days. Did the tremor make her nervous or was it just normal behavior? The geese in the other flock have not started sitting yet. I have been impressed by their frantic eating during the last weeks. They have been devouring grass as if obsessed with it. Every time I have had to confine a flock to a small holding pen in order to be left alone while working in that part of the vineyard, several geese have managed to fly back into the pasture to eat. The ganders have not been too concerned about feeding, and have stayed in place content with just watching the females.

Two days ago the mother goose started sitting on the eggs again and has not left once. The book says thirty days until the goslings are hatched.

Today we got the boulder we have wanted for so long. It is a two-ton stone from the Stone Hill quarry outside Molalla, fairly flat and wide, shaped somewhat like a fat rune stone, but not as smooth. Flecks of greenish mica still sparkle from the newly broken surfaces. The side that was exposed to the weather in the quarry is a rusty grey.

The man with the backhoe places it standing up, erect like a raised hand. He buries the base about a foot and a half, leaving a little over five feet above ground. I make him align the thin side with the north-south layout of the gardens and the buildings even though two tons of rock is hard to maneuver and he keeps asking: "Is that good enough?"

"This is a good boulder," Carl says, "Look, it has this place where I can put my foot and climb on top of it." As he talks he quickly shows how to do it. All afternoon the boys play on it.

Later, when I find an almost perfectly round stone the size of a basketball, I place it on top. For weeks this stone creature startles us. Cynthia tills a twelve-foot circle around it. This year she will add some compost and plant potatoes, later divide the circle into quadrants with the paths aligned to the cardinal points. This small circular garden will be like a compass, reminding us of the four directions.

Bud break on a grape vine is when the swollen, fuzzy buds open and the first leaves become visible. At this point the leaves are very small, the size, they like to say, of mouse ears. In our vineyard this normally occurs during the first week of April, but this spring everything is a week late due to that heavy snow and cold weather at the end of February. It could be a problem if there are further delays later on, since ripening grapes here can be marginal for many varieties. But there is nothing we can do about the weather except observe and speculate about the coming vintage, and that is exactly what we do.

When we bought this land, it was just a big pasture where a local rancher ran his cattle. We were later told that he was the only one in Beavercreek to use a horse to get around and check things out, or to round up his herd when they had to be moved. But cattle were becoming unprofitable and many pastures were planted with Christmas trees; others were put up for sale.

When the rancher took us out to see the land, he drove us in his big pickup truck, cattle scattering in all directions as we bounced across the hard ground. He drove us to the top of the knoll to show us, not the view of the rising and falling contour of the Cascade ridges in the east, nor that of the valley spreading in the west below us with the pale coast mountains far beyond it, but instead the small view to the north, in an opening between two groves of tall Douglas firs: downtown Portland. Yes, there they were, the high-rise buildings glinting in the distance, houses spreading in all directions in the smoggy air, the radio masts on the West Hills above it, the city we had left less than an hour ago. "Look at this view," he said confidently, "this is where I would build a house."

On my mother's birthday I notice that the bleeding hearts are in full bloom. It was one of her favorite flowers; planted in a shady corner of my parent's garden it was visible from the front porch of their summer cottage. At their latitude it bloomed in June. She called it lieutenant's heart—courage, passion, and good looks combined, hinting at her weakness for fancy uniforms.

She was thirteen years old when she left her small village just below the Arctic Circle to help an older sister who had moved to the city and needed a nanny for her first child. She never returned home. Instead, she pushed on from that provincial town to the capital, drawn to its lights and its multitudes. Whatever she knew of farm life, she kept to herself. The unbroken, 120-year-old chain of our family farm traditions snapped. And before I even understood that this had happened, all I wanted was to get out of the city and back to the land.

Early this winter a friend gave us a bleeding heart with completely white flowers, and now it is the very first to bloom. The flower has exactly the same shape as the regular bleeding hearts, but there is no pigment, no blood, only the pure emptiness of white. I can see it from the front porch.

A solitary V of wild geese flew north overhead, honking for the stragglers to catch up. I kept my neck tilted until they disappeared behind the trees, thinking how seldom I see them, but how often they flew across the pages of the local poetry anthologies of the 1930s. Geese, salmon, old-growth timber, indicators of ecosystems, traveling together on the road to silence while the media talks about growth, jobs, and a healthy economy.

All the canes from the vineyard that came from the pruning have been left in a huge pile. It has been a favorite hangout for frogs and Oregon juncos, but it is time to get rid of it. The easiest thing would be to burn it. That would be cheap too. But I have started mulching with wood chips in so many places that I am always in need of more, so I rent a big chipper in Oregon City and spend a day transforming the vines into chips.

Spread thickly enough, chips are excellent for preserving soil moisture and suppressing weeds, because they are themselves weed-free, compact, and fairly durable. Whatever grows through is easily removed. I use them around newly planted fruit trees or shrubs, on flower beds, or on paths in the garden. On the ground, fungi and micro organisms work on breaking them down. Worms consume what they can. And as they are converted into humus, I simply add more on top. It is just an accelerated process of what happens on any forest floor.

A couple of weeks after the Oregon juncos have departed for the mountains, the goldfinches suddenly announce their presence. They fly together in colorful flocks of a dozen birds or so, eating the dandelion seeds. The first time I glimpsed one, I thought it was an escaped cage bird, maybe a yellow budgie. The colors seemed far too bright and exotic for a bird this far north.

Soon after their arrival Nils finds a goldfinch head on the ground. The decapitation has been very precisely and cleanly done, and the head is added to his treasure box, next to the obsidian arrowhead, the snake skin, the abalone shell, the rooster leg bone with the spur, and the lump of fool's gold.

I catch a small headline tucked away in the morning paper: FARMERS ON FOOD STAMPS. The piece below is brief, simply quotes some census statistic on the number of recipients, which has gone up. There are no explanations.

I think to myself: if life got really tough in one way or another, would our family be in that group too? Or could we feed ourselves on this eight-acre farm? We don't grow any grains, so if we had to do it there would be no bread or cereal. We have no animals which give milk, so no butter or cheese. What would we eat? Potatoes, beans, vegetables, corn, fruits, juices, eggs, chicken, goose, lamb. And wine, we could drink plenty of wine. I think we could stay alive, but our meals would be rather simple. Still, the idea intrigues me.

"Cynthia," I say, even though she is deep into another section of the paper and I should know better, but the headline has fired up my imagination and I can't help it, "what about trying to grow all our food for one year, be totally self-sufficient? I have read that with bio-intensive methods four thousand square feet of garden space will feed one adult a complete vegetarian diet for one year. What a challenge, Cynthia. We won't buy anything. I don't mean next year, or even the year after that, just any year. We could plan it, grow some grains. We could even get a milking goat, or a cow. And we could have meat—chicken, goose, lamb—since we have pastures."

"Are you crazy, you blue-eyed Swede? I like having a cup of coffee. I like pasta. I like a piece of cheese. It would turn the pleasure of gardening into a nightmare of necessity. Totally unrealistic, and just like you, too."

"Not forever, just for one year," I try, but I know I am speaking to deaf ears now, "one year is only a percent or two of our lives, what's that? What a satisfaction it would be."

"Not for me," she rumbles from the depths of the newspaper.

I look at the garden through the window. Except for a profusion of asparagus spears, there is not very much to eat out there right now, and if supplies were low, it would be hard. Maybe that is why April has been called the "cruellest month."

This time of year the clouds are often a mix of white and very dark grey, moving quickly with patches of deep blue sky in between. Below, the light changes swiftly. The mood of the landscape swings from cheerful to gloomy within minutes. One moment the sun shines with steam rising off the ground, the next it turns dark with April's sweet showers coming down in sheets.

I am out by the wood pile when it suddenly begins to hail. I have never seen it come down as heavy and hard as this before, and take cover under the lean-to. The hailstones are as big as fresh, green peas bouncing off the earth and rolling into place. I can hear their low drumbeat on my neighbor's plastic greenhouses. I can see the hail joyously dancing on top like a confused chorus line before sliding off with a fast rustling. It hits the metal roof on the chicken coop with a loud, steady roar. White drifts collect on the roof and noisily slide in the downspouts. The grass bends and receives them without a whisper. The kiwi gets its tiny leaves pierced. In the sky, millions of white spots draw trajectories against the black clouds. Then it ceases as suddenly as it began, the ground white and silent. A few minutes later the sun is out.

There is a call from the chicken hatchery: our two dozen Orpingtons hatched out OK, but the vaccination machine is broken and they cannot vaccinate against Marek's disease.

"Do you still want the chicks?" the woman from the hatchery asks. "You don't have to take them if you don't want to. I can have another batch ready in ten days, maybe two weeks, and by then the machine should be fixed."

We start looking at the schedule. The way the chicken coop is set up, the first flock needs to be old enough to roost under cover outside before the meat chickens arrive five weeks later. That way we do not have to mix them and will eliminate pecking problems. A two-week delay? How likely is Marek's disease anyway? We have both grown tired of the silence and decide to go ahead.

While Cynthia drives down to the hatchery to pick them up, I spread a layer of chipped grape vines on the floor, rig up a heat lamp, put out food and water. When she unloads the day-old chicks from their little box, she dips their beaks in the drinking fountain to make them aware of the presence of water. They are very perky and run around full of energy.

One morning a couple of days later I find one dead on its side right under the heat lamp. One leg is black and shriveled, as if atrophied. The others have pecked out its black eye ball and there is only the socket left. I leave the dead chick outside on the ground for the cats, but they will not touch it, and in the evening I have to bury it in the vineyard. The body is only about the size of a sparrow or finch.

As spring accelerates towards its full crescendo we start planting potatoes. The ones that have been stored in the coolest corner of the wine cellar have good sprouts and are ready for the soil. Cynthia makes furrows and I fill them with wheelbarrow after wheelbarrow of sifted compost. Then the boys help Cynthia plant the seed potatoes in the compost, while I write down the number of potatoes per row, carefully keeping track of the different varieties before the soil is raked back over: yellow Finn, Yukon gold, Mandel, Charlotte, Sangré, Åke Truedsson blue, Rödbrokig Svensk, Rote Erstling. They are fingerlings and oblongs, early and late maturing, boilers, mashers, and fryers.

To get a true sense of what a variety is like, one should grow them for at least a couple of seasons. Potatoes could be a lifelong undertaking. It has taken us a long time just to track down some of these heirloom varieties, but now we have a garden full of old Swedes returned home to the New World.

From time to time towards the end of the month all of us start visiting the mother goose, looking for goslings. I admire her stamina. I often wonder what she broods about. What does she know? Does she worry about the weather? Where she will take the goslings? Or does she already know everything she needs to know, and simply waits content with the rain, the sun, the wind, and the stars? Sometimes a week will have gone by without me observing her leaving her nest, but occasionally she will get up, run slowly for a while, stop and flap her wings, do her yoga exercises—stretching the left leg backward and the right wing forward and vice versa—maybe preen a bit, and absentmindedly eat some grass. Her legs have turned very pale.

One morning Cynthia finds the two ganders outside the mother goose's nesting box, talking and being very protective. Cynthia has to put a piece of plywood next to the fence to avoid getting bitten as she observes them. She tells us that she heard the high pitched voice of a gosling from under the goose, but nothing was visible. In the late afternoon, a small yellow head emerges from under the mother's wing, looking at us for the first time. The next morning there are two goslings, which, by the end of the day, start showing signs of being impatient with the nest. The following morning six more have arrived, and later in the day the flock leaves the nest, never to return.

Nils and Carl find six abandoned eggs in the nest. In one of the eggs they discover an almost fully formed, stillborn gosling, in another, a partially formed fetus. The other four had never been fertilized.

In the evenings I read Konrad Lorenz' studies of greylag geese, the ancestor of all the European strains of domesticated geese. I marvel at the nuances and richness of his observations, and begin to glimpse the world that geese live in. Three thousand years of domestication have eliminated the desire to migrate and increased their size, but their flock behavior is more or less unchanged. I think of us humans, who, before we became settled, were sophisticated ethologists for so many long millennia. Standing in the light rain, watching the movements and interactions of the new flock, the hunter and gatherer in me is schooling an aboriginal eye.

On the last evening of April, back on the other side of the planet, they are celebrating the ancient custom of *Valborgsmässoafton*. In anticipation of spring they will light big bonfires to welcome it, choirs will sing in its honor, politicians will give speeches, poets will recite poems, the crowd will dream of summer in the icy wind, and there will be the drunks, the lovers, the old people and the children, the fireworks, the stars, the flying embers, the pile of coals lasting well into the following day. The people will have repeated the ritual of renewal for another year.

Here, after receiving an invitation in the mail from the local Swedish-American Cultural Heritage Organization for the coming Saturday, which is not the last day of April, to celebrate this event (misspelled, like almost everything else these second and third generation "Swedes" attempt to write in Swedish) by having a potluck salad in the basement of Norse Hall, I think about the definition of "cultural heritage."

Would it have any meaning even if the custom was carried out properly, since the larger culture does not recognize what it is? Does it need that kind of affirmation? How much context does a piece of "cultural heritage" require to become meaningful? Does it matter that almost no one in that organization speaks or reads Swedish and that their meetings are always carried on in American English? That they do not regularly follow contemporary developments in Swedish politics and culture? That they do not really know what *Valborgsmässoafton* is?

Putting the flyer away I make up my mind about not renewing my membership and not attending any more of their events. I may not fully know who I am, but I know where I came from and where I live, and I refuse to pretend anything else.

Five

Inspired by Kerouac's On the Road, *and limited by my meager traveling funds, I hitchhiked from Montreal to Los Angeles, experiencing the gradual unfolding of the enormous vastness of the North American continent. Its size was far beyond anything I had ever been able to comprehend from my road atlas. Most of the cities I saw left me cold, but there was a certain quality about the land, especially in the arid West, that I instantly grew fond of; or maybe it was just the absence of human activity that appealed to me.*

Having arrived on the West Coast, I soon made my way north to Santa Cruz where Cynthia was finishing up her studies at the university. It was late fall by the time I arrived and she would be done by early summer. I felt like a messenger who had crossed half the earth just to deliver one question: When you are done here, will you return with me to that old farm in the middle of the forest? Because if you will, I could wait here and we could return together. For some inscrutable reason, she answered yes.

It was the first winter in my life without snow; December was something like a cold, wet Swedish summer. I walked around experiencing everyday life, read a lot, contemplated small mysteries like the "PED X-ING" painted on certain streets, or what they sold or did in a "body shop." And then there were those garbled words the café waiter uttered when I ordered my cheese sandwich for lunch. It probably took fifty sandwiches to realize that I had always answered "yes, please" to his inquiry: "Would you like mustard or mayonnaise?" I, who came from a city where only drunks or crazies addressed strangers with anything but a question for directions, discovered that Californians struck up conversations with strangers about everything. And I would walk along the shore and never tire of watching the surfers trying to catch the great waves that the winter storms kept rolling at the beaches.

For nine months I thought I was waiting to go back home, when in fact, I had begun to lose what I thought was home.

As soon as the pinot noir vines start growing they are susceptible to powdery mildew, and to fight it, the organic grower must spray the emerging leaves every ten to twelve days with a sulfur and water mix. This must be continued until the grapes start changing color in August. I have come to dislike spraying very much. It seems that no matter how carefully the sprays are timed, the grapes always get some powdery mildew anyway. And when they do, they spoil so badly that they become useless for wine. The sulfur also acidifies our already acid soil, requiring additional liming. It is work leading to more work, expense begetting expense. And why spray when there are mildew-resistant varieties?

Considering the fact that no one in the Willamette Valley can grow the traditional European grape varieties without constant spraying, I began to wonder how it had been possible to grow healthy grapes to make good wine before sprayers were invented. Why did farmers in centuries past bother with grapes if powdery mildew inevitably ruined them? It took some reading to find out, but the answer was surprisingly simple: prior to the 1840s, Europe did not have powdery mildew; the disease was brought from America. For European viticulture, the New World turned out to be a Pandora's box. Native American vines, imported as curiosities into the botanical gardens of Europe, carried not only powdery mildew, downy mildew, and black rot, but the phylloxera root louse as well. Ironically, the grape vines from the New World also turned out to be the saviors of the European vineyards, when it was discovered that the traditional varieties could be grafted onto phylloxera resistant, American rootstock without much noticeable loss of quality. But growing grapes suddenly became a much more costly and labor-intensive undertaking, and the rootstock provided no protection from the mildew. Even grafted vines had to be sprayed.

The consequences of fighting foliar diseases on grapes have been costly. In many places, after a century of spraying a blend of copper salts and sulfur known as the "Bordeaux mix" to control the downy mildew, toxic levels of copper have built up in the soil, making it unfit for food production. Few talk about the cost to future generations of our present agricultural practices, and what a hundred years of accumulated synthetic fungicides, pesticides, and herbicides might add up to.

Swallows love to fly. I like to watch the violet-green swallows swoop and arc in the empty space over the vineyard. They relish the exhilaration of speed and scream with delight. Sometimes they fly in tandem, one following the other, copying all its loops and turns until they suddenly part, only to converge moments later. Sometimes they fly up in front of the bird houses to look them over, and eventually they will peek inside. If they like what they see, they will start gathering nest material, mostly goose feathers and grass. They will pick up a goose feather from the ground without landing, then arch back into the sky. Sometimes they let the feather go, or drop it, only to swing back down to pick it up in mid-air. This may happen several times, and it is hard to tell whether they do it for fun or just to get a better grip. And how do they know that a feather is good to use for a nest?

When they fly for joy, they do not seem to feed. The feeding flight consists of shorter dips and twists; it has more interruptions. Feeding is a solitary kind of flight. Flying together is for love.

As Cynthia and I walk through the flowering orchard a gun shot cuts through the early evening. It is rapidly followed by another, and another until we lose track of the number of shots. As we hurry back up to the ridge to try to see what is going on, there is a brief pause. We can discover neither shooter nor target. There is no sign of anyone when the firing starts again, but this time some switch has been flipped to automatic and the shots come in bursts so fast they cannot be counted. Then nothing. The statement is clearly made and a bird sounds a single, hesitant note in the deep silence that follows.

Our original flock of geese has grown to seventeen birds. The mother goose leads her goslings around the pasture looking for tender grass and leaves while the ganders keep a watchful eye.

They grow very fast. At night, and during rain, the mother shields them under her wings. The adults talk to the young ones, keep them together, show them how to preen. The only tense moments occur when the goslings slip through the field fence, leaving the worried adults on the other side. The heavy geese, normally so content with walking, will gather their strength and fly over the fence to join their young. Occasionally one of us will find a goose on the wrong side of the fence, forlorn and upset, too tired to follow when the roving goslings have once again slipped through the fence to join the flock.

Even though Konrad, the dominant male, clearly is in charge in this flock, the second in command, Uncle Ernie, assumes the role of parenting if the mother and father disappear too far. The cohesiveness of their extended family is very strong.

Today, it has been exactly seven years since I began my weather record, recording maximum and minimum temperatures, rainfall and snowfall, fog, hail, and strong winds. Every day I have transcribed the temperatures in a monthly graph, curves rising and falling against the established Portland averages, which are the closest statistics available. I fit three years on one piece of standard size graph paper. What wild oscillations between warm and cold! I am beginning to understand the meaning of an average: it is an abstract line that reality crosses on its way from one extreme to another.

Just as I pull up with the tractor to park it, a young rabbit takes off from a tuft of grass in front of the wheels. It is the size of a kitten and it makes for cover among some rocks close by. Like a child covering its eyes thinking it is invisible, the young rabbit has just managed to hide its head. I can't resist the urge to pick it up. It is the first wild rabbit I have ever caught with my hands. It is soft and beautiful and the boys beg me to put it in a cage, and I do.

An hour later, ignoring the water, fresh grass, and leaves they provide, it has not stopped throwing itself against the wire cage. When it finally hurls itself against the metal wires hard enough to split its lip and start bleeding, even the boys are eager to release it.

When we eventually built our house, we did not put it where the rancher had suggested; we planted vines there and let them have the view of the city. Instead, we choose a place further back on the ridge, close to a corner of the property, with tall firs behind us and a view out over the vineyard and the forest and mountains beyond. Along the top of the ridge the land is fairly level and well drained, and none of the gardens require any terracing. It is easy to get around. From here, we can follow the sun's uninterrupted arch from dawn to dusk.

The exterior is that of a typical Swedish house, simple enough in its lines and proportions to lead people to ask if we have renovated an old farmhouse. I even imported a couple of barrels of traditional Swedish red stain for the board-and-batten walls, and painted all the window trim white as tradition requires. Seen at close range against our background of fir trees, the picture seems very authentic and could have been taken in any number of Swedish provinces. But it is only an appearance; the interior is not at all traditional. Instead of the conventional house full of small rooms separated by hallways and passages, and a kitchen spacious enough for a typical three-generation household plus farm hands, we wanted the spaces inside open and connected; a house full of air and light that fit the needs of the four of us and our way of life.

Late in the day, after three days of warm, still weather, the wind suddenly starts to pick up out of the southwest. It arrives with a rare, violent intensity that rips things loose and hurries them along the ground. An empty plastic milk jug disappears into the vineyard. Pieces of paper whirl by. Two pieces of corrugated metal roofing take off like a pair of big wings from the compost bins. The aluminum drain pipe hums a low vibration against the house. The young, pliable trees bend and nod like heads in an audience listening to a convincing argument. It smells like the ocean again, cold and clammy. We hurry and put the tools away. When the rain hits, everything gets soaked instantly, as if the raindrops were larger and wetter this time of year.

The energy is quickly spent. Two hours and eight millimeters later, there are still grey curtains of rain in the mountains, but here, only scattered clouds and a late evening sun. Leaves and plants saturate the air with sweetness. The slant of light and the long shadows are Nordic and melancholy, and my whole being suddenly aches for what is no longer home.

Some time after dawn we always see crows flying west out of the foothills, down into the valley, commuting to work. A few stop in our vineyard almost every day, and if one of us runs the tractor, more will come. During the day they will comb the vineyard floor in their well-worn tuxedos. One will act as a guard on a post while the others scavenge whatever calories they can get. They are easily spooked and will not let me come close. They think a stick is a gun. All I have to do is show a stick to them, and they will take off, but only so far. As soon as they are out of range they will land again and go about their business. I thought of building myself a Norwegian crow trap and trying to catch them for food, just like in the old country. But it never came to more than a thought. I like the intelligent glint of their eyes; they add life to these acres.

In the early evening, they come flying back a few at a time or in large flocks. The day's work in the valley is done. They talk to each other, perhaps complaining about hard times among spreading subdivisions taking over farmland, exchanging news, or just telling jokes. Further up in the hills, where patches of second-growth fir form the fringes of the mountain forest, we see them circle together, cawing, getting ready to pick their roost.

At the end of a warm day I notice a strange-looking animal far down our driveway. It is sitting perfectly still, watching me, and I start walking toward it. It looks like a puppy, but as I get closer I become less certain. The shape is very odd. My mind starts its process of elimination: It is not a cat, because the shape is wrong; not a dog either, and definitely not a possum. Maybe a raccoon? At thirty yards the shape splits apart, and three kittens scurry back toward the gate. When I get down to the road I find a pink plastic hospital tray with some crumbs of cat food left in it, and a knocked-over container that had held water. The terrified kittens are piled in the ditch between the fence and the road, trembling and meowing.

I walk back up to the house to get a box to put them in. They scratch and bite as I put them in the box. One even defecates on my pants as I lift it up. I lock them into the garden shed and block the cat door, much to the irritation of our own cats. From the noise they make, it is obvious that they will not accept the kittens. We really do not want any more cats, so we talk about what to do. The Humane Society is an hour away and wants a ten-dollar donation per kitten, and they will put them to sleep if nobody adopts them, but it is the best solution we can come up with.

In the morning, as we get ready to take them, we find that they have escaped and are now hiding inside some blackberry brambles on the other side of the fence. The kittens are still so frightened that they will not let themselves be picked up voluntarily, so we enlist the help of our neighbor's teenage son to trap them. He spends the day with Nils and Carl rigging snares and luring them out with food. They catch two. The third disappears deeper into the woods never to be seen again.

The buzz of hummingbirds has become a daily occurrence throughout the flower beds and the garden. Cynthia and I rig up a feeder under the lean-to roof and notice that they come to feed there in the evenings, loading up sugar for the night. We sit down with the boys and tell them to be absolutely quiet and still. It is a difficult task. But our waiting pays off: a hummingbird swoops down and drinks; then, instead of leaving, alights on a rusty wire coil nearby to watch the dying light. The radiant colors on its throat shimmer as if alive, but what is even more astonishing to me are the minute mechanisms of the legs and feet clasping the wire with such immaculate precision.

One afternoon when the locust trees are in bloom, a walking salesman arrives. His hair and clothes are wet from the recent shower and his whole appearance is odd. He is not carrying a sample case either, just a small plastic bag. Before Cynthia has a chance to tell him that she is not interested in buying anything from anyone today, he draws a few lines on his faded jeans with a black ballpoint pen. He will now demonstrate how easily this is removed with the cleaning fluid he is selling. But before he has a chance to get the cleaning fluid out of his plastic bag, he catches Cynthia's persistent "No, I don't want to buy anything." Surprisingly, he drops his sales pitch immediately and asks where the nearest neighbor lives. It is a superfluous question since the house next door is clearly visible. Cynthia shows him the gate in the fence and the man takes off. I walk back on the driveway to see if a car is parked down by the road, but the road is empty.

A few days later Cynthia runs into the woman next door and asks her what she thought of the itinerant salesman. The neighbor woman is surprised. No salesman ever came to her house and, as Cynthia learns later, nor to any of the other neighbors.

Not all ridges divide watersheds, but ours does. The water from the roof of our house is picked up in underground pipes and led to a small pond for the geese. From there it seeps down the hill to join the water from a small spring in the gulch below. Eventually it hits the south fork of Beaver Creek, which shares the ravine with the old railroad grade. Beaver Creek then runs westward until it empties into the Willamette River above Oregon City's Willamette Falls. But when we walk out on the east slope of the vineyard we enter another watershed. Here, all the water rambles south down the steep hill, then enters the north fork of Buckner Creek, which drains into Milk Creek, which drains into the Molalla River, which then empties into the Willamette a couple of miles upstream of the mouth of Beaver Creek. And only a few miles higher up into the foothills to the northeast, all the creeks drain into the Clackamas River.

It is a long ridge and it does not surprise me that the old Indian trail went up on it. Once it might have been the dividing line between the Clackamas tribe to the north, and the Molallas to the south, an old line of demarcation as forgotten as the Molalla tongue spoken here 150 years ago.

The blacktail deer have returned to our hill. Among the Christmas trees on the gentle rise to the southwest Cynthia spots four of them, standing still with their big ears turned toward us. Through the binoculars they look like two does and two yearlings. No one can discover any fawns. After a while they continue west along the ridge, probably because there is nothing to eat. Three Mexican workers with backpack sprayers just went through applying herbicide, and now there is nothing but browning weeds, sickly looking moss, bare dirt, and small firs.

I think the deer were eyeing the west side of our vineyard, where the vines have grown rapidly and there is good cover and plenty to eat. They like grape leaves very much. It would not be the first year they moved in.

After a whole afternoon of mowing the section of the vineyard where there are no geese or sheep, I finally park the tractor on the highest point, kill the engine, and pull out my ear plugs. In the silence that follows I look at all the cut grass and the words come back to me: *today there are more people alive on earth than all those generations combined who preceded us.*

Our Orpingtons have feathered out and are, at four weeks of age, old enough to spend the nights under cover outside. Their space is needed for the meat chickens, and the small door to the inside of the chicken coop is closed. I put on a respirator and shovel the dry, dusty litter into a couple of empty feed sacks. There is at least enough manure to power up a good cubic yard of compost. Then I hose everything down and leave it to dry. The next day I spread out fresh, dry wood shavings, get food and water ready, check the temperature under the brooder lamp.

The day-old Cornish cross chicks that Cynthia brings home from the hatchery look just like the others did a month ago, and so another life cycle is initiated. The weak will cull themselves out, like the one whose anus burst with guts spilling outside the body, or those who mysteriously lose energy and simply lie down to die. I bury four out of twenty-six before the dying stops.

My father writes that his mother has died. She had just turned ninety-eight. Her small, frail body has been transformed into mineral ash and buried next to my grandfather in the cemetery on top of the hill. In my mind I return to the white church whose name I have forgotten. There, it should be the time between the flowering of the bird-cherries and the lilacs, the light of the long northern evenings.

In May, my grandmother would take her basket and pick the emerging tips of the stinging nettle. She had a couple of places she would go to, and she would pick them without gloves. It took me a long time to trust her assurances that they did not sting when they were young and tender. What I wanted to know was this: when did they start stinging? How could she tell that they were still safe to touch?

When she had a basket full, the two of us would walk home and she would steam them for soup. It was a simple soup. Just some butter, flour, and milk, to which broth, garlic, and the steamed, chopped nettles were added. It was served with finely minced chives and diced hard-boiled eggs, and tasted intensely like spring. Today, having no nettles, I made a similar soup using kale instead, salvaging the last leaves before the plants started blooming. Squatting in the garden picking the curly green leaves into a basket, I thought of her, and how our lives are repeated in such endless variations.

Six

We arrived at the communal farm in Sweden early in the summer, and soon realized that there was trouble in utopia. Nothing had been planted or prepared, and we had arrived too late in the season to do anything. But there was a more serious problem. The farm was located fairly far north; the elevation was high, and the growing season turned out to be short, with less than ninety frost-free days on the average. We learned that it had snowed on June tenth one year. We had done our homework poorly—this was not good farm country. But then we were not farmers either, but young idealists from the city dreaming about rural bliss. All we knew was how to talk about bio-dynamic principles; no one knew anything about how to actually farm.

We discovered that most of the villagers did not know what to think of the long-haired people who did not speak their dialect. Waiting to find out, they had adopted a strategy of guarded reservation. It did not take us long to figure out that the area was ruled by timber companies, and that the locals, whose grandparents or great-grandparents had sold off their forests a century ago, had become proletarians reduced to competing among themselves for the scant jobs that were left. The village was small, the clearcuts large, and both the school and the store had closed years before we arrived. We lived in the village, but we were an island.

It also turned out that the commitment of the people in the commune to the original dream was subject to revision, and that few wanted to live permanently in a village where nothing happened. Most continued to come and go with the ebb and flow of money. Mixed with all the good intentions of sharing things in life were the insolvable complications arising from different personalities residing under one roof. There were unvoiced complaints, division into different camps, and an annoying absence of straight talk, all of which slowly began to corrode our vision.

Cynthia and I eked out a living for a while before we finally gave up and took the train back to Stockholm to make some money.

In the middle of turning a compost pile, while marveling at how well our discarded, Indonesian reed laundry hamper has decomposed in just a few months, and thinking about how the garbage problem would disappear if all household objects could either be recycled or transformed into compost, I find my gaze unexpectedly seized by the green curvature of the land. I stop working and just stare at it. I am riveted to its sensuous shape, which suggests the softness of a body, even though it is simply cropped grass with rows of vines. The green of the grass is electric with presence and energy, and the dark trunks seem caught by surprise, frozen in the middle of some mysterious motion. Yet there is no event taking place in front of my eyes, not even any birds moving. As I take this in, I become aware of myself watching.

I think it was James Lovelock who suggested that all sentient beings might simply be the eyes of the earth watching itself, an evolved extension of the earth's own consciousness, the planet becoming aware of itself, and all our observations and emotions a kind of reporting back. That was what I felt like: eyes working for Gaia.

Grape vines will not bloom in overcast weather. The flower buds can wait fully formed for a very long time. Cynthia and I learned this one year when a grey May was followed by a cloudy June. It was neither particularly cold nor rainy, just persistently overcast. There were three scattered days of sun each month, and that was not enough to convince the buds to open. When the grapes finally did bloom it was not until July, a month late, and the grapes never caught up and had to be harvested unripe when summer finally collapsed toward the middle of October.

But when the sun comes, the buds will unveil the inconspicuous flowers, each a tiny crown with a row of white antennae. Bloom is hesitant at first, with a few scattered flowers here and there in the clusters, then turning into a surging wave of froth. No bees need to pollinate the flowers; they are serviced by the wind. If the evening is calm and dry, one can smell a thin jasmine scent. It is a critical time for the crop. If it starts raining during the peak of bloom, pollination will not be complete and the fruit will be uneven.

This year it does not start well. Just as the Maréchal Foch starts flowering the weather turns rainy, and the forecast calls for an extended period of overcast and showers. But we are lucky. After three days of intermittent light rain and clouds the weather turns again, becoming sunny and warm. A few weeks later bloom is over. The clusters are formed, the new, green grapes the size of pinheads.

Cynthia is awakened in the wee hours of the morning by a strange scream. She slips out of bed and enters the boys' room to listen. No nightmares or fevers toss their small sleeping bodies, just normal quiet breathing. When she returns I wake up too. What was it? She does not know, except that it had the sound of a child crying in his sleep. The silence lingers. We get back under the covers, and as our bodies search out their most comfortable positions and we start drifting back to sleep, the eerie cry is repeated in the white night. We both recognize it this time: it is a young cockerel, trying out its first adolescent urges against the paling sky.

When people we know, who live in cities far away, learn that we have a vineyard and make wine for the family from our own grapes, strange seeds are planted in their imaginations. Always beyond our control, they grow into notions of what such a life is really like. "Ahh, like Falcon Crest," they say, raising both eyebrows simultaneously and smiling, knowing fully well that it is not like that at all, yet retaining something they think must apply. Or they superimpose memories from journeys they have made through landscapes with vineyards, or add recollections of small Mediterranean harvest festivals or tasting tours in the California wine country. It is meaningless to respond with facts. Our small acreage and odd grape varieties are abstractions that mean nothing.

Many volunteer their hands and feet to help pick and crush the harvest, and, of course, to drink the results; some decide to come and take a look. A few are even surprised that we are not harvesting grapes when they arrive, or that harvest is not an ongoing thing all summer: "But I see grapes in the supermarket all the time!" Having come this far, they want to see the vineyard. We show them, and it is like walking into silence: there we are in the middle of the intense sunlight, just rows of leaves moving in the breeze, the grapes green and tiny and months away from being ripe. What is there to say? That we are in the middle of removing the suckers that sprout from the lower part of the trunk, that there are literally thousands of vines, that the work is tedious and that we get dirty and sore from all the bending down? A vineyard trembling in the heat does not usually hold any visitor's interest for very long.

The wine cellar under the winery is more appealing. It is cool and dark. We bring some crackers and cheese with the basket of glasses and the wine thief, the long tube with which we siphon wine out of barrels and carboys. Then we start tasting our way through the varieties and the vintages, talking about flavors and aromas, the importance of oak cooperage. We explain the annual cycles of the wine through the barrels and the bottling process, talk about the increasingly acknowledged health benefits of a moderate consumption of red wine, about dishes that would make good matches to the wines we taste, and about how to try for harmony between the garden and the cellar.

School is out and the boys put up a big tent in the yard. It will be their summer residence, and it takes them a whole day to fill it with the things they want: mattresses, pillows, sleeping bags, quilts, chairs, a table, a chest of drawers, books, water bottles, an extension cord for lights and a radio. They are not going to be roughing it. The cats get a place too, with water and food.

Without realizing it, they have placed the lamp low at the far end of the tent, and every night after dusk their shadows are cast on the tent cloth. Whatever they do becomes a shadow play for us to watch. What a delightful and unexpected form of parental surveillance!

One evening Nils is alone in the tent, on his bed reading, when Cynthia notices a silhouette. "Look," she says, "you can even see the cat eating." As I look I realize that the size of the body and the wide tail, with its characteristic bushy, question mark shape, is not that of a cat, but of a skunk. But before I have a chance to open my mouth Nils has discovered it too. He has dropped his book and is sitting perfectly petrified on the bed. We debate what to do, whether to interfere or not, but decide to trust luck. The skunk will not be rushed and takes his time to finish off the cat food before eventually departing, trailing his potent odor with him into the darkness of the vineyard.

After several days of training young vines in the replanted section of the vineyard, my new work boots are finally broken in. The stiff leather has softened and shaped itself to the irregularities of my feet, and I don't have to put on any more bandages for blisters. At the end of each day I leave them on the porch to air out, but one morning a boot is missing. Cynthia and the boys help me search around the house and the yard. Whenever a gate has been left open during the night, roaming dogs sometimes visit the compost piles, and there is one in particular, with hairs reddish in color, that I suspect. I walk up and down the driveway, looking down the rows in the vineyard, then head up the main road. Just beyond the curve on top of the hill I run into two neatly dressed middle-aged women out walking, one holding a small spray can of tear gas, and without explaining the whole story, I ask them if they by any chance have seen a new boot along the road. They look at my unshaven face, then at my feet, perhaps to check if a boot is missing. But no, I have a shoe on each foot. I can sense their minds trying to figure out how and why a person looking like me could have lost a boot along the road. No, they haven't seen a boot today.

A few days later I resign myself to the fact that it is gone, and drive back into town to exchange sixty more dollars for another week of blisters.

For years I wondered why so many French grape breeders had worked on developing hybrid grapes to achieve a combination of disease resistance and good wine quality, and then never had their achievements promoted very much. The names of the new varieties are rarely seen on wine labels and very few people know much about them. Our main grape, Maréchal Foch, is a good example. Even though it was developed by Eugene Kuhlmann in the Alsace region in the 1880s, it is virtually unknown in France today. When I have asked people traveling to France to inquire about them, they have returned empty-handed. It has always been very mysterious.

Then a small group of people from Europe on their way to a permaculture conference in Eugene stopped by the vineyard, and among them was René. He had an answer. It turned out that he was very well informed about the issues concerning the hybrid grapes versus the traditional varieties, which he insisted on referring to as the "noble varieties." René had worked as a lobbyist for the French wine industry in Brussels, where some branch of the European Union regulatory commission was responsible for establishing rules for the internal EU wine trade. The lobbyists had, among other things, succeeded in making it illegal to use the word "quality," or any of its synonyms, with all wines made from hybrid grapes. They had also managed to make it mandatory that such wines carry the words "table wine" on the label. The implications for the price structure of wines from the different grape categories were not difficult to grasp.

"The argument is quite simple," René said. "You know, all the hybrid grapes are crosses between noble varieties and other members of the Vitis family. Now, because all these other grapes have a different chemical composition of sugars, the hybrid grapes inherit them. And when that is fermented into alcohol, well, the alcohol becomes different too. And that, *mon ami*, is where the important point lies!"

Here René paused for a dramatic effect he had obviously practiced in Brussels on several occasions, his dark, intense eyes penetrating my incredulous gaze. I was not sure what the point was.

"That different kind of alcohol," he slowly continued while nodding his head, "makes people mad when they drink it! Mad! It is that simple."

Confronting a French syllogism of this magnitude I was uncertain as to what I could say. Admitting that I had drunk the wine from a hybrid grape for almost a decade would be as close to an admission of madness as one would ever come, and it would naturally disqualify whatever argument I would make. I was trapped, as I was supposed to be, and had to admit that it was an interesting theory.

Later, watching them leave, I was glad to be where I was, well beyond the reach of the powerful interests of the entrenched European wine industry. I realized that at least in some respects, I was still living in the land of the free.

The meat chickens have almost four weeks of rapid growth behind them now, and, even though they do not know it, have already arrived at the midpoint of their brief careers. Their white plumage is patchy, with reddish-pink flesh showing through in several places. Their legs are bright yellow and their feet unnaturally large. Several have deformed toes. Bred for a brief life in a small cage, the entire bird is beginning to look like something that natural selection would never have permitted.

Up until now they have behaved pretty much like any young chicks, but the result of their insatiable appetite is beginning to show. They stand up on wobbly legs, take a few stumbling steps, then stop and sway, as if unsure of the power of their own legs to propel them forward. Unconvinced, they sit down again, preferably right in front of the feeder. Only very exciting events, like the boys throwing them a handful of fresh worms, or the low-flying silhouette of a hawk, will induce them to run for more than a couple of feet. This spectacle of growth is both impressive and alarming at the same time.

June is always a busy month for the hayers in Beavercreek. The weather is carefully watched and when a solid stretch of dry, warm weather is predicted, everyone jumps into action. Tractors with haying equipment drive by throughout the day, and the drivers are dusty and rednecked. If I am down in the vineyard working I always wave, and some return my greeting; others don't. I don't know any of them by name. Every hour of daylight is used as they move from field to field turning the drying hay into long, fluffy lines. The air is filled with the sweet smell of hay. Once dry, the balers crawl over the stubble spitting out the pale green, abstract cubes, with pickup trucks, trailers, and field hands following.

Every year, yet another hay field or two are replanted with Christmas trees or lost to subdivisions. Some of the remaining fields will still be hayed, but most of the small ones are left to themselves and expanding patches of blackberries.

Our large flock of geese have eaten everything they like down to the nubs, and are actively reaching for any drooping grape cane. Some vines are hit pretty hard and it becomes urgent to get the geese into another pasture with fresh grass. Unfortunately, the gate to the new pasture is not in a corner, but in the middle of a fence line, and it is very difficult to get the geese to move through an opening that has not previously existed. Even though the gate is fully open they will not go through; in their minds it is still impassable because it has always been so. It makes me wonder what preconceived notions do to one's view of the world. What new openings am I myself blind to?

Cynthia and the boys come to help me herd them through. We try to be slow and gentle, but still they keep running right by. The third time around they get trapped behind the gate and absolute chaos erupts with the entire flock climbing on top of each other, honking and flapping their wings in panic and desperation. As they scatter, a goose with a broken leg remains. It is one of the young ones born in the spring. She tries to join the flock by pulling herself forward, using her wings like oars.

As I approach to pick her up, Konrad comes charging. It is impossible to tell whether he is actually attacking me or the goose. Perhaps he really is attacking the hurt one, because as soon as I pick it up he turns back. We will not even attempt to set the leg since we intend to eliminate this flock in the fall anyway, so we put it in a wire cage with a bowl of water. There, it will spend the night before it is slaughtered. This time of year it will have very little meat, no fat, and a nice liver. Cynthia will save the down.

On the day of the summer solstice, I notice a large work crew along the road putting in an optical fiber cable. It is a quickly moving operation with a truck carrying a big orange spool of cable, a backhoe, several gravel trucks coming and going, and two flaggers with stop signs and walkie-talkies directing the traffic.

I ask one of the flaggers what they are up to, and learn that they are installing the information superhighway. I am not opposed to technical innovations—I drive a car; I like working on a word processor; I'm lured by the speed and low cost of sending e-mail around the world. But here the word "highway" implies an openness and a freedom which I am not convinced is there. What will come through these cables? Where will I be able to go? Who decides about the map, the content—he with the most money? I wonder what he will decide. A neighbor talks about interactive TV with great enthusiasm, but isn't it interactive only in the euphemistic sense that one could say that shopping is interactive? I can live without having every international corporation in the world displaying their wares inside my house. I don't need more, less would be just fine. Somehow I suspect that this cable is just another hookup to some giant electronic Coliseum, a colorful river of images difficult to verify, a reality devoid of smell, taste, or substance.

I walk back to the house and get a large bowl from the kitchen before I disappear into the garden. This is the season of abundance: though the strawberries are almost gone, the blueberries have just started to ripen, and the raspberries are in full swing. The wind plays with my shirt as I pick what has ripened since yesterday. The raspberries are so sweet and juicy, and my mind drifts to jam and syrup, pies and desserts.

After dinner Nils and I decide to walk over to the nearest cemetery, called Ten O'Clock Hill Pioneer Cemetery, just a stone's throw up the slope from the Ten O'Clock church. The name comes from the fact that the congregation did not have the funds to buy a real clock for the steeple once the small church was built, so in the interim they painted a clock face showing ten o'clock. When the real timepiece was eventually installed it was already too late. Ten O'Clock was the name that stuck, and it now designates not only the church and the cemetery, but the long hill and the area around it.

The place is only a mile away from our house, and I must have driven past it hundreds of times, yet I have never stopped before. We discover that the first burial, placed in the corner of the graveyard nearest the main road, took place in 1883, then the rows follow straight and neat as the years move on. Not even a third of the total cemetery is yet filled. The early names are almost all German: Moehnke, Steiner, Schwein, Staub, Mueller, Schweizer, Brunner, Bluhm, Mau, Schwichtenberg, Schumacher, Weiler, Schmidt, Swartz, Bittner, Landeck. On several of the early gravestones even the epitaphs are in German: "*Ruhe im frieden,*" "*Ruhe im Herrn,*" "*Ruhe sanft.*" This strong German past in an area I have come to think of as mine is news to me, even though I should have realized it from the names echoed on some of the local roads. Our road leads to Spengler Hill, and was, I assume, named Spengler Road before it became Spangler.

The continuation of the early settler culture is as absent as that of the Molallas. No one speaks German here any more. The old homesteads don't even look particularly German, and there are no festivals or fairs celebrating vestiges of ethnicity. Here, everything unique and alien has been absorbed and transformed into that powerful current that is mainstream American culture; nothing seems capable of resisting it. Like a meat grinder, it eventually makes cultural hamburger out of us all.

As we walk home from the cemetery picking up empty pop cans and beer bottles for the five-cent deposit, Nils asks if I want to be buried at the Ten O'Clock Cemetery, raising all those questions I sometimes ask myself but never answer. "No," I say, "if I'm going to be buried in Oregon, I think I'd like to be buried on our farm."

"Are you going to write something on your gravestone?"

"I don't think I want a gravestone," I say, "but I haven't decided that yet. I'll think about it and let you know."

"That's OK," he says, "I'm going to think about what to write on my gravestone."

But the conversation meanders onto other topics, and I never find out what he wants to write. I don't ask. We make sixty-five cents in cans and bottles before we head up our driveway.

Seven

After leaving the farm, the dream of communal living in the country quickly vanished. It was obvious even to us that we had been naive beyond belief, and it was embarrassing to think about it. So we didn't. Instead, we found a place to live in the city, found work, reconnected with old friends, and started saving for a trip through Europe. Half a year later we were ready to take off. We traveled for eight months with backpacks and sleeping bags, visiting practically every single country of Western Europe, and ended up drifting through the Greek archipelago for a couple of months without any particular goal, taking each day as it came.

I had thought of this journey as a planned exploration of various places and sights I had always wanted to see, but it turned out to be more of a haphazard odyssey of random experiences. We moved outward through Europe, but kept arriving at ourselves. One day we would run out of money again, and then what? We were traveling, but we were also waiting for a decision. What were we going to do next, or, more specifically, what was I going to do next? There was nothing waiting for us at the farm. We could return to the city, find a job again, and set out on another journey, but wouldn't it be just a postponement?

Eventually I realized that what I wanted most was a formal structuring of my ongoing interest in American culture, and I was ready to go to school to get it. Back in Stockholm, I discovered that the Swedish university system made students wade through what seemed like too many semesters of required grammar and language before they let them get to the literature. Again I found myself in the wrong place. Cynthia was ready to return home too, and we started making preparations for moving back. After gathering and submitting a thick pile of official documents to the U.S. embassy, I was granted an immigrant visa to the U.S.A.

Alone in the evening, turning the pages of a glossy wine magazine left behind by an elderly visitor, I get depressed by the images associated with wine. No people can be detected in these magazine vineyards, and when the wine is drunk, it is not by families gathered around the daily dinner table enjoying the gift of all that is grown. There is not a single calloused hand pouring a glass in the ritual of simplicity and friendship.

Instead, every picture appears carefully designed to trigger an envy reflex: elegant European environments, beautiful young couples, fashionable dress, unlimited expense accounts, the life of the rich and famous. The lead story talks about Bordeaux futures and the ins and outs of safe wine investments, perhaps the inevitable result of a society where less than five percent of the population farm for the other ninety-five percent.

The young vines we planted in the spring show signs of thirst, and I spend several sun-baked afternoons in a wide-brimmed hat and long-sleeved shirt rolling out drip-irrigation tubing and attaching emitters to each plant. What pleasure it is to finally open the faucet at the end of the week and let the water run through the tubing. As the steady trickle hits the dry brown earth it crackles and foams like the froth on a fresh espresso. The young vines respond within hours, the luster returning to the muted leaves; within a week they are growing rapidly again. From now on, they will receive at least four gallons every two weeks through the middle of August.

I walk among the rows listening to the water trickling out, making sure that none of the emitters are clogged with insects or dirt, studying the shapes of the leaves of the new varieties, admiring their vigor and energy, and wondering what the mature grape clusters will look like. The plants feel like children and I look forward to watching them grow. A glass of wine is still four or five years away.

Beavercreek is known as a rocky area, yet there are very few rock structures. There are few stone houses and no rock walls. Perhaps most people find the stones too round to do anything with them, but for a couple of years I have been building a long rock wall parallel to the driveway. It is slow work, but it is enjoyable and uses up a lot of stones. I quickly ran out of what we had found in the vineyard and garden, and ended up buying a truck load. Now that is gone too, and I am collecting stones with my tractor and trailer from a neighbor down the road. Laying the mostly roundish, red rocks into something resembling a straight wall takes time, since each stone must be studied and properly turned before it is placed. Often it takes several tries before one finds the right fit. I have noticed that some days it is easily done, every stone is always the right one. At other times no stone seems to fit; it is very much like writing.

It has occurred to me that my stone wall serves no practical purpose, that it does not enclose anything. It is just a straight line two feet wide and four feet high, open at both ends. When I am done with this wall along the driveway, I could turn and make a corner, and then continue for twice as long. If I did, it would take me years and still not enclose anything. With stones one can create the illusion of permanence and order, hold back the fluidity of nature, and put the mind at ease. There are few things that are as abstract as such stone walls.

In the cellar, the pinot gris finally seems to have completed all the changes I have been waiting for. There are no signs of fermentation any more. A thin, yellowish layer of sediment covers the bottom of each carboy, but the wine is clear. Smelling it, the wine has a clean, flowery nose with none of the yeasty aromas that were there in the spring. It reminds me of pineapple. The flavor is crisp, reminiscent of citrus, perhaps not as elegant and smooth as it could be, but definitely without major flaws and certainly drinkable.

To get a clear wine, the sediment must be eliminated prior to bottling. To accomplish that, I lift each of the carboys onto a sturdy shelf above a meticulously cleaned stainless steel barrel. No matter how carefully I lift the carboys, the sediment is stirred up, requiring another week or two of settling.

Moving them I wonder about the word "carboy," why American English came to prefer it to the word demijohn, versions of which is what most European languages seem to use. The sensuous associations of the shape of "Dame Jeanne" rejected in favor of youth and machines? As a four-year-old, Nils was fascinated by the word and always used to ask me about the carboy: "But who drives, papa?" Was it the boy who drove the car, or the car who drove the boy? And how do you answer that?

The soil we farm is called "Jory;" it consists of a foot or so of brown topsoil, then a varying depth of reddish clay with pieces of soft conglomerate rock deeper down. There are some boulders scattered in the ground too. The soils in these foothills were "formed in colluvium," meaning that they are the accumulations from the erosion higher up in the Cascade mountains. The soil is characterized as having two limiting factors—low fertility and relatively high acidity.

Vines do not require high fertility, but they do like a more neutral pH, so I am slowly liming the vineyard to neutralize the pH. An advantage with Jory, some claim, is that the high clay content is able to retain as much as ten inches of the winter rains into the summer, and this provides moisture to the deep-rooted vines after the rains have ceased. Others argue that really good wine is made from vines grown on more well-drained soils, where stress is common and the grape size smaller. It is all part of that elusive *terroir*, the character of the soil showing itself in the wine.

Below the topsoils, the well report journeys through layers of lava like a geological time log: clay brown, lava grey soft, lava grey fractured, lava grey hard, lava grey and brown, lava brown, lava brown soft, lava grey soft, clay yellow. The yellow clay is 155 feet below the surface and marks the bottom of our well. My neighbors north and south on the ridge use water from the same aquifer. As I contemplate the well report I notice that some of these layers of lava are thicker than the height of our three-story house. We are living in an area of active volcanoes.

By this time most of the grape buyers have already contacted the growers and reached agreements for the coming harvest. The wineries start early, the home winemakers late. One winery is always first. The winemaker makes up his mind in the early spring, and we settle on quantities and price. He buys Maréchal Foch from five different growers, but claims that the wine from our vineyard stands out among them because it has a bittersweet dark chocolate flavor added behind the fruit. Occasionally, I can taste that flavor too, but not having that range of vineyards to compare with, it is not always so easy to single out. It could be the result of our low crop levels, which concentrates the flavor, or it could be our organic cultivation methods. It could also be the unique marriage of our particular iron-rich soil and the Maréchal Foch grape, the flavor of the land showing itself in the wine.

The home winemakers are usually slower to decide if they are going to buy our grapes or not. Some, who have bought from us in previous years, keep coming back, but there is always a rising tide of calls as harvest approaches. We have already sold out our entire estimated harvest and have started putting people on a waiting list. Maybe there are more grapes than we think. Estimating yield is one of the most difficult things for us, since the vineyard is still so young. Once we sold a ton more than we actually had, and everyone grumbled and had to do with less. Ever since, our estimates have been on the conservative side.

There are many different kinds of contracts between growers and buyers. When we first got started, a winery offered us a long-term agreement for our entire harvest, which would eliminate the necessity of having to market the grapes each year. The winemaker offered us a low base price per ton, plus a bottle price determined by the retail price of the wine. We would bring the grapes to the winery, but get paid nothing at delivery. In January we would get the first third of the base price, in April the second, and in July the remainder. We would get paid the first installment of the bottle price when half of the inventory was sold, and the rest when the last bottles had been sold. It would take either grower or winery three years to terminate the contract. We considered this proposal at length, but finally rejected it as much too limiting, leaving us too dependent on a single buyer. We preferred to sell our grapes on the open market, though in some ways this is riskier.

I have heard it said that each cell in the body is replaced every seven years, that it is continually renewed from within. Now our bodies are rebuilding themselves out of carrots and beets, green onions and Swiss chard, with plenty of garlic added. Like most things one can grow, there are innumerable varieties of garlic. From an older woman who took a liking to Cynthia's garden, Cynthia received a handful of bulbs of her favorite varieties: red Russian, white Italian, French rocombole, hot Korean, silverskin, and one called the Safeway special. They are certainly distinct varieties, but who knows what their true names are. My favorite is the French rocombole, because even though the heads are small, the cloves are large and the thick, purple skin peels easily.

    Wave after wave of shelling peas reach their peak of firm perfection, and we do our best trying to harvest them all. I like picking them in the morning before it gets too hot, then sit with Cynthia in the shade on the patio talking, shelling them into a bowl in my lap, tossing the empty pods into a bucket. She blanches and freezes some of them for later; others find their way into a creamy pea soup with fresh tarragon and butter dumplings.

When the boys sit in the garden and shell the sugary peas straight into their mouths, tossing the pods over their shoulders, losing track of time and of themselves, I glimpse the light of the plants glinting in their eyes. They do not know it yet, but it is a kind of takeover from inside.

The purple lavender starts flowering. Women gather at the neighbor's herb farm for an annual event to make all kinds of lavender things: wands, arrangements of sprigs, miniature baskets, small sachets for the dried flowers, scented things to put in drawers and closets among handkerchiefs and dress shirts. No men are allowed at this event. Several times I find that I stop in the middle of working with something, and just stand there, listening to the wind carrying the faint voices and the laughter rising and falling from the shade under the dark fir trees. I can't make out any distinct words. I recognize it like one recognizes a dream, in which I sail below the rocky cliffs of Lesbos, knowing that some part of me has heard these sounds for thousands of years.

The meat chickens have grown plump and heavy, and have become almost completely stationary. They hardly move when one of us comes in to add food or change the drinking water. Only one more has died since they first arrived, leaving twenty-one survivors. In the other partition of the chicken coop the Orpington cockerels have started fighting, and most of them will also have to be eliminated. As far as we can tell, there are eleven of them, and we decide to keep the three best-looking ones for the time being. We will let them fight it out and put on some weight, then cull two of them later. All in all, there are twenty-nine to dispatch, and we decide to do it in two days.

The day before the first slaughter we put fifteen of them in cages with water, but no food, so they will empty themselves out. This way, a slip of the knife will not release the contents of the gut into the cavity of the bird. In the early dawn we gather what is needed for the slaughter: a stainless steel bucket for the scalding, a thermometer for getting the water temperature right, a waste bucket for heads, feathers, legs, and viscera, a knife and a sharpening stone, and a cooler with ice. While Cynthia starts heating water I get the axe and chopping block ready. I grab the first chicken, wrap its wings and body tight in a burlap sack, say a brief thanksgiving and apologize for taking its life before I bring the axe down. All fifteen bodies are bled, scalded, plucked, cleaned, and refrigerated to an increasingly intense hum of flies and wasps. Slaughtering after a good frost in the fall would be nicer. We save the livers for paté, the necks, hearts, and gizzards for broth. We are done around noon.

Is it worth it? They cost fifty-five cents apiece, and ate two and a half sacks of feed at twelve and a half dollars a sack. We also bought a new waterer for twelve dollars, but it will last at least another year. Cleaned out, each of the meat chickens weighs about three and a half pounds, and the cockerels are not quite two pounds. We know what they ate. They were given no antibiotics, medication, or hormones. They were free to run around to the best of their ability. They lived as a flock within the natural cycles of light and dark.

The gophers are multiplying in the vineyard and brown piles of fresh dirt appear everywhere. Due to the rather wet spring there seems to be plenty of food, and since they have not eaten a single vine yet, I am lazy about trapping them. The boys get two dollars a head, and Carl is an expert trapper, already well beyond twenty this year. Before breakfast Cynthia and I see him disappear out from the tent to check his traps, and sometimes he returns shaking a dead gopher at us through the window to confirm that we owe him two more dollars.

One gopher finds his way into the garden, straight into the potato patch as if guided by a compass, and instantly the situation becomes critical. Ignored, he could eat or remove every single potato. Piles of damp, dark soil show where he has started cleaning out under the plants. He has struck gold and he knows it. Coming in perpendicular to the rows, he seems to have sampled the different varieties until he got to the seventh row with Sangré, deciding that was the flavor he was after. Here his tunnel turns ninety degrees, straight north, following the mother lode. As I dig along this tunnel only a few scattered potatoes remain. Some are half eaten; some have chew marks in them. The tunnel runs about a foot down in the ground, exactly in the middle of the row. Small vertical openings at regular intervals show his technique. This gopher is a potato pro. Eventually, two thirds down the row and after twelve feet of exposed tunnel, I come to a dead end. From here on, the potatoes come in thick.

I return to where I first started digging and continue in the other direction. Here, the tunnel twists back and forth, then dives below a large stone at the edge of the garden. Eight feet of tunnel in this direction. The stone is huge and it takes half an hour just to get it loose, then I have to get the tractor to pull it out with chains and drag it to my stone wall along the driveway. Where the tunnel disappears out from the garden I set the trap. If I am lucky, he will be dead within hours.

With all the surface moisture gone, there is little growth on the vineyard floor. The new weeds that manage to grow are ones that geese do not like. The sunlight is savage and sharp. Ribbons of light flutter among the leaves as they are rustled by gusts of hot air. The color of the vineyard has changed from the light green of spring to the dark green of mature leaves. The growing tips of the canes dry up and fall off, and the wood starts turning hard, changing from green to yellow and brown.

It is the turning point in the growth cycle of the vine. Up until now the canopy has been primary, but from now on the energy will be directed into the fruit. The small, perfectly shaped grape clusters have swollen considerably, but they are still tight and firm and intensely green, far from the days of *veraison*, the moment grapes show their true color.

I am working by myself down in the new vineyard anchoring the irrigation tubes between the weed fabric and the mulch, then heaping the mulch back over the tubing. By keeping the black tubing away from direct sun I hope to keep the irrigation water a bit cooler. Suddenly I am aware of crickets singing all around me, but when did they start? Did they always sing like this? I listen hard, and it sounds as if many of them have synchronized their sounds, a rhythmical *see-saw, see-saw, see-saw*. Then behind it, another group, also in unison, comes in with a counterpoint: *eeh, eeh, eeh, eeh*, fast and repetitive. Are they consciously creating this symphonic sound or is the pattern just in my imagination?

In the middle of listening, I become aware of a tight column of spinning wind moving slowly and diagonally down the vineyard in my direction, lifting dust and small pieces of dry leaves and grass. I have never seen anything like it. Why, on several acres of land, does it come right at me? The column is perhaps six or seven feet tall, two feet in diameter, and charged with an extraordinary presence. It passes so close that I could take a step forward and reach out and touch it, but I am frozen to the ground, goose bumps running down my back. I cannot move until it disappears into the old vineyard on the other side of the driveway. I walk over there, but the whirling column has dissolved into thin air. In the evening I return to the words attributed to Chief Seattle:

> "And when the last Red Man shall have perished, and the memory of my tribe shall have become a myth among the white men, these shores will swarm with the invisible dead of my tribe, and when your children's children think themselves alone in the field, the store, the shop, upon the highway, or in the silence of the pathless woods, they will not be alone. In all the earth there is no place dedicated to solitude. At night when the streets of your cities and villages are silent and you think them deserted, they will throng with the returning hosts that once filled them and still love this beautiful land. The White Man will never be alone.
>
> Let him be just and deal kindly with my people, for the dead are not powerless. Dead—I say? There is no death. Only a change of worlds."

Eight

We settled in Oregon and I enrolled at the university, and Cynthia found a job that paid the rent and put food on the table. I signed up for as many American literature classes as I could. For years I lived with my eyes glued to the pages of books, enjoying every day of it. My hunger for literature was insatiable. There was so much to discover, so much to learn! I was introduced to different traditions of writing, and was taught different ways of reading a book. Patterns emerged, and I had the feeling that the structures of understanding I had been wishing for began to take shape, and I discovered that obtaining an undergraduate degree was not the conclusion of my undertaking, but the beginning.

I had been gone from Sweden for over three years and I had an education, but no particular vocation. I was still not qualified to do anything except possibly read and talk about books, and that was not something employers were looking for. I felt restless again. I had become homesick for Sweden. During my whole stay I had been cut off from Swedish news, out of touch with current events back home, and I felt that this had begun to erode my sense of identity. My family was still there and I missed my old friends. I missed the language, the snowy winters, and the light summer nights. In my mind it all grew into a sense of home I could not experience in the United States, which, when I raised my eyes from the book I was reading, seemed too self-absorbed, too mercantile, and only too eager to make a joke out of everything. I wanted to move back to Sweden, and Cynthia did not mind. Leaving would not be very difficult—there were just the two of us, with few possessions to hold onto. After a few months we were ready to leave.

A heat wave drives us outside at night. The inside of the house refuses to cool down, and I bring a mattress out onto the stone patio. There is a light breeze from the east. Even though Cynthia hosed everything off after dinner, the stones are already dry again. They are still warm to the touch. The house too radiates heat back into the night. Our cats are overjoyed that we have come to join them and cannot stop purring.

From our patio we can follow the seemingly endless stream of airplanes approaching Portland International from the south, bright lights shining under their wings. There is an occasional distant roar from a semi on the highway, and for a while we can hear sirens on Beavercreek Road urgently hurrying through the darkness disappearing down Ten O'Clock Hill. As the vehicle disappears, we hear instead the faint rock-and-roll from a radio somewhere to the west, and voices of children playing at the neighbor's. Deep silence, like that of the desert or the back country, rarely comes here.

Lilies, nicotiana, and night-blooming stock perfume the darkness. The silhouettes of a few bats crisscross above with a kind of nervous eagerness. Further up, a brief silver line streaks against the blackness of space.

In the early morning, lively sounds of salsa music from a transistor radio across the road drift in over the vineyard. Farm workers are shearing the neighbor's Christmas trees. In the upper section of the field, where the trees are big enough to be harvested this fall, I can see the men slowly circling the trees, trimming them into perfect cones. Their Spanish voices laugh as they swing the long, sharp machetes. I like how it brings that otherwise empty field to life.

After lunch they are either tired of the music or out of batteries, because the radio is quiet. The heat has taken its toll too, and there is little talking or joking. Every one is working barechested and silent. As I come down to the mailbox to collect the day's mail, all I can hear is the slow, rhythmical *swish, swish, swish* of the shearing. Some have rolled bandannas into headbands to soak up the sweat, others have tied them into a kind of cap with a knot in each corner. Focused on their task, they do not bother to greet me. I am impressed that they still have the strength to swing the machetes at all; my arm would have given up a long time ago.

There is an intense fragrance of Douglas fir in the air, and it lingers for days. Once, having returned to Stockholm from traveling in Greece, I called a state-run wine information hot line to ask where the resinous flavor of retsina came from. Was it some particular grape variety that had that unique flavor, I wondered. Without hesitating the man at the other end answered no, and began explaining how the Greek vineyards were almost always in the vicinity of pine trees, and how the hot Mediterranean summers sweated such quantities of pitch from the trees that the aroma itself was transferred onto the grapes and from there got into the wine. Hiking around the Greek countryside Cynthia and I had both seen a lot of pine forests next to vineyards and often smelled the pitchy fragrance in the air, so the answer sounded plausible enough. I thanked him and hung up. It never occurred to me to ask why no Greek red wines had that resin flavor, or why not all Greek wines tasted like resin.

Now, knowing what I know of the ancient Greeks coating the insides of amphoras with pitch to make the terra cotta wine-tight, thus starting a two-thousand-year-old tradition of flavoring wine with resin, I wonder what other strange notions about wine are still being perpetuated.

On the west side, two deer keep coming into the vineyard to feed on the grape leaves. They are a buck and a doe. Every time I come running they quickly turn back and jump the seven-foot fence, but today the doe is a bit too casual and barely clears the top wire. A big tuft of hairs with a smear of blood on it is still vibrating on a barb when I get there. I can see them on the other side of the fence along the edge of the small Christmas trees. They have slowed to a gentle trot, lowered their tails and turned their heads looking back at me. "You watch out," I call, "hunting season opens pretty soon. I'd be careful if I were you!" But they have seen and heard me so many times during the summer that this is nothing they worry about. Ignoring me, they demonstrate their lack of concern by stopping to nibble some weeds along the edge of the field.

In the cellar the pinot gris has cleared again, and all I have to do is carefully siphon the wine into the steel barrel below. By consolidating the wine, any variation in taste between the different carboys will be eliminated. Having it in one barrel will also make continuous bottling much easier. The siphoning must be carefully done: if the sediment is agitated, the wine will be cloudy. Carboy after carboy silently slips through the clear tubing and the flowery smell of pinot gris fills the cellar.

The next day we are ready to bottle. The wine is now siphoned from the barrel into the bottle filler, where Nils and Carl take turns hooking on empty bottles and passing the full ones to me. I operate the corker and Cynthia is in charge of moving bottles, keeping everyone supplied and preventing collisions. There are no mishaps and no broken glass. A wine barrel holds about 275 bottles of wine, an impressive sight, as pleasing to the eye as a bin full of potatoes or the long shelves of canned goods in the larder.

Once the bottling equipment is rinsed off and put away, I start up the labeller. The labels have been printed with the vintage year and trimmed to size. I feed them through the old machine, stick them onto the bottles, and Cynthia and the boys adjust and straighten. The job is quickly done and we return to the house in time for lunch, letting the glue dry overnight before moving the bottles to the cellar.

Newly bottled wine goes through something called bottle shock. From its restful existence during fall, winter, and summer, it is rapidly put through a series of transfers and mixings. Somehow, all this agitation within a short period of time makes newly bottled wine taste rough and sharp, even a bit jagged. It needs time to calm down and return to its former self. How long that takes seems to depend on what type of wine it is: only the unprejudiced palate will know when a wine is ready to drink.

About a quarter of a mile away, on the north side of the ridge where the road curves through a small forest, someone has dropped two roosters off. Being out of view from any house, the place is often used to dump garbage and kittens. Early in the mornings I hear the two new roosters crowing and our cockerels answering. The hens listen too. In a week the roosters have worked their way close to our vineyard, perhaps attracted by the response. The neighbor's teenage boy has discovered them as well and comes over to ask me if I want the roosters if he shoots them. I say sure, if you help me clean them, we'll have all of you over for *coq au vin*.

Later in the day he brings the roosters over, trying to be casual about his hunt among the Christmas trees. Several times he tells me in great detail how he, while running, shot one and how it died instantly. We clean the birds together. Nils and Carl come over to watch the scalding and the plucking. Cutting one of the birds I carefully separate the crop from the esophagus and analyze its content: an abundance of shiny greenish black beetles and pieces of windfall apples from the old wild apple tree along the road.

Every summer the Oregon State Fair organizes a statewide amateur wine competition. It is very popular. Over a hundred wines are entered, red and white table wines according to grape variety and style, as well as sparkling, fruit, and berry wines.

Each winemaker submits two bottles for each category, one for the tasters, the other for display during the fair. A panel of five judges evaluate the wines, and each bottle will get five brief, anonymous notes.

Most years I enter a wine or two, and sometimes I have won nothing. I have only received a first place once, for our 1992 Maréchal Foch, when the wine was summed up in the following way:

> "Deep purple; berry fruity; round, soft, voluptuous finish; a very pleasant easy to drink wine."

> "Dk purple; blackberry, spice; rich, round fruit, med. tannins."

> "Very pretty, dark; good, forward; dark cherry, medium tannins, very nice."

> "Beautiful deep burgundy; wood spices!; bitter herbs, dark fruits; very inky, nice!"

> "Soft, jammy, yum; *Very* good wine! The only one I swallowed."

We have salads for lunch, and dinners are anything we feel like cooking from the cornucopia outside. There is so much food coming out of the garden that we cannot eat it all. Cynthia often works late into the evenings preparing things for freezing or canning, putting up food for the winter. And even though most home-grown food smells, tastes, and stores better than what any supermarket sells, it reaches beyond fragrance or taste, beyond minerals or vitamins. A garden puts eating into a familiar context. The loops of energy are small and local. It makes things clear and connects us to the ground. Everyone in the family lends a hand growing it, seeing it change from seed to food. We observe the power of water and compost. When plantings fail, it teaches humility. A garden is consumption preceded by creation, a kind of self-reliance that is a doubling of one's enjoyment.

Then it starts, that first sure sign of the approaching harvest which we have been waiting for—the coloring of the grapes. It is as if small purple light bulbs turn on, just individual berries here and there at first, then a great wave of purple sweeping through the green clusters. In another month they will look almost black. After a week, most of the Maréchal Foch has changed color. The pinot noir starts turning a little later, but the event is overshadowed by the visible signs of powdery mildew on the grapes. In spite of eight conscientious applications of elemental sulphur, it is a fact that we will not harvest any of the grapes. As I walk through the remaining section of pinot noir, looking at the wasted fruit, I realize that this is not my grape.

Cynthia and I have not grown grapes long enough to know when the turning should occur to be called normal, especially since the weather seems to have been somewhat capricious during the last decade. Now the turning happens before it did in 1991—a very late year for everyone—and after it did in 1992, a very early year for everyone. Perhaps it means that this is closer to the average.

One day Carl finds the first egg from the Orpingtons. He spots it in the dust in the middle of the chicken yard, as if dropped there by accident. It is not very big, light brown in color, and just over an ounce in weight. When I look up egg weights in a grading table, I discover that it belongs to the smallest of the official egg sizes, the size smaller than "small" referred to as "peewee."

Three days have gone by since bow-hunting season for deer opened, and for a week there has not been a trace of the buck and doe that spent so much time in the west vineyard. Then today, under a grey sky, from one of the upstairs windows I glimpse the familiar brown bodies among the green leaves. Adrenalin surges through my body, transforming it. I become my father and grandfather, my uncles and cousins; I become the men of the northern forests before them. I am simply repeating thousands of years of hunting moose and deer, fox and hare, and this hunter decides what to do. I slip into my shoes, grab my bow and arrows and sneak around the corner of the house, making sure I get as much cover as possible. When I come around the raspberries I discover that the deer have moved into the more open pinot noir, and as I see them, they see me. I stop and they stop eating. We look at each other and they decide to retreat into the dense vegetation of the Maréchal Foch. They move slowly and deliberately, but without lifting their tails.

As they disappear among the green leaves I decide to wait for a minute, calm myself, and get an arrow ready on the bow. As my heartbeat quiets down I move forward again until I reach the first row. Cautiously I peek down the row—no, they are not there. I walk one more row—it's empty too. The third row is the same. Just as I begin to think that they have headed all the way out of the vineyard and disappeared among the Christmas trees as they do when they are chased, I find them in the fourth row, oblivious to me now, eating leisurely on the grape leaves. Both animals present perfect broadsides. The buck is turned to the west and the doe is eating on the row to the east. Their heads are gone among the leaves, otherwise the entire bodies are visible. There is no indication that they worry about me. But they are far away, perhaps two-thirds down the row. As I position the bow I realize that it is just like field practice where you can stand up straight and take your time. I try to calculate the distance in my head. Twenty-five vines in a full row, eight feet between each vine, two-thirds of a row down. My mind draws a blank. Forty yards maybe? My sight is set at thirty yards so I know I have to compensate by aiming high. But how high? I decide to trust my luck. I draw the string, aim just over the back straight above the heart. I am surprised at how calm the hunter is; then he releases the arrow.

Before I even have a chance to lower the bow I see the buck collapsing on its rear legs, front legs kicking, trying to pull the rear up again. But the entire back half of the body seems paralyzed, and as I watch I can see him drag himself in under the grape canopy and then remain there, still. The doe walks up to the buck, their heads are very close, as if they are going to kiss each other. It is now that it happens, that soundless, telepathic communication I often sense between deer. The doe says goodbye, then turns and runs in a direction opposite the one they always ran together before. I will not see her again this season.

I sit down, put my bow to the side, and let the minutes tick away to make sure the wounded limbs will stiffen. Time dissolves into a river of now. Eventually I approach, walking slowly. The head is turned away from me. I stop ten yards behind it, and I can see the arrow sticking out of the upper part of the rear haunch. The arrow really moved sideways in its trajectory; I must have twisted my arm left or the bow string right at the moment I released the arrow. A little bit of blood has trickled out of the wound. The buck is not looking at me but I know that he is aware of my presence. I retreat, fetch a sledgehammer from the tool shed. This time he turns and looks at me. I stand behind him, say a small prayer out loud, thanking him for the meat and for letting me shoot him. I apologize for having to take his life. All this, I realize later, I said in Swedish. Then I strike him hard between the antlers, and life gurgles out of his mouth and the eyes go dim.

With some effort I manage to get him into a wheelbarrow and roll him down to the winery. The body has been transformed into something resembling a leather sack full of liquid. He is heavy; I can barely hang him up by myself. A light drizzle has started falling, washing the blood into the grass. As the body cools, the ticks start leaving the fur, and I discover two larvae, one in each nostril, also trying to leave. I wonder what they are. Since this is the third deer I have ever killed, I still need to consult the manual for field cleaning. I work for three hours, bury the guts, the lower part of the legs, head and broken skin—I am still not very good at skinning—then hang the carcass in the cool wine cellar overnight. I am completely soaked through when I am done.

It takes four hours to butcher it the following day. I follow the instructions in "Boning Out Your Deer" from the local Extension Service office, which describes the cuts traditionally done by the Native Americans. It is the first time I use this guide, and the body comes apart in immaculate pieces: two shoulder roasts, two arm roasts, three back-strap pieces, one tenderloin, two ribs, two sirloin tips, two rounds, and nineteen pounds of ground meat. Cynthia makes venison soup stock from the bones, which she cans, and the chickens get to pick the boiled bones clean.

In the late afternoon air the intense aroma of the wild Himalayan blackberries down behind the house wafts all the way up to the patio, luring us down to them. How well the fruit makes its shining blackness known. I hope it will keep every bird belly so happy and full that they will not care about the grapes in the vineyard, but I know it is just a hope. By the time the grapes are perfectly ripe, the blackberries will be long gone.

After dinner the boys and I go down to the patch behind the house to pick enough for a pie. We gorge ourselves too and return with scratched arms and stained fingers and lips, planning tomorrow's attack: an early start, a couple of small pails, thick gloves, loppers, scraps of plywood. We will pick for blackberry jam, and Cynthia promises us a big blackberry cobbler too.

A friend unexpectedly stops by late in the afternoon. He is on a bike ride but ends up staying for a venison dinner with salad, mashed potatoes, steamed green beans, carrots in a nutmeg *sauce velouté*. I uncork more wine than usual to celebrate his surprise visit, the venison in the freezer, and the coming harvest. When the evening is over, we worry a bit about his departure. As he has no light for his bicycle, the boys lend him their small squeezo flash light, which he clenches between his teeth as he straddles his bike and takes off. We wave and cheer as his pale halo of light disappears down the dark hill.

Nine

Cynthia and I arrived in Sweden in early summer, and the light was as beautiful as I remembered it. We visited family and friends, even made a trip back to see what had become of the old communal farm. All the communes in that part of the province had disintegrated, and almost everyone had returned to the cities. Where we had lived, only one of the original couples remained. They did not farm the land and kept no animals, but she had managed to get a job in a school nearby and he was handy, a mechanic by nature and training who had also become a skilled carpenter. They had children now and were part of the village, and their life seemed so simple and graceful compared to the restlessness we had known.

In the fall I was in graduate school, buried in the library, and again Cynthia had found a job that would pay the rent and put food on the table. It should have been perfect, since I had returned home to all the things I had missed. As the months went by, it slowly came to me that in spite of having come back, I did not feel at home. Even though most things looked the same, they seemed to have changed. Swedes seemed too reserved, too formal, too pessimistic, too xenophobic, and it suddenly bothered me in a way it never had before. Eventually I realized that it was not Sweden that had changed, but me.

At a New Year's party, a tipsy civil servant, who had pontificated at great length on the merits of the Swedish system, concluded by asking Cynthia why it was that, whereas she spoke Swedish with an American accent, he did not speak English with a Swedish accent? He said it in earnest, even though his accent was as thick as porridge and his grammar dotted with errors, and suddenly he seemed the perfect symbol of what was wrong: It was a smug and paternalistic society, where an overly self-confident government bureaucracy, in its well-meaning attempt to make everyone feel secure and safe, was much too engaged in regulating people's lives and making decisions for them.

On Friday at five o'clock the Department of Environmental Quality closes its office and turns on the answering machine. A courteous message tells you that the office will open again Monday morning, and that you may register any air pollution complaints on the answering machine. I suspect that every grain and grass-seed grower, every slash-pile logger, every backyard garbage burner knows this. This afternoon it is the grain growers. Just around five o'clock a couple of distant white columns rise into the southern sky like fat arms waving. After dinner the entire sky has a white tinge, rusty red in the west, and all through the evening and night the sharp smell of burnt straw permeates the house.

A boy I have not seen around before, maybe eleven or twelve years old, rides his bike down the main road at least half a dozen times. Maybe his family is visiting someone at the bottom of the hill. I see him walking all the way up to the top of the ridge, red in the face and with his helmet dangling from the handlebar. It is a warm day to walk up a steep hill. I wave to him as he walks past, and he waves back to me, but neither of us says anything. He has his helmet on when he comes down, leaning forward, pedaling hard, going very fast. As I follow the drip irrigation lines to check that the emitters are not clogged, I think of Sisyphus and all the meaningless things we do, but then I remember the boy's smile and my thoughts change direction. After that, the rest of the day is easy.

Walking down to get the newspaper right after dawn, Cynthia spots a dead goose at the distant end of the pasture. The flock is agitated and far away from the white, lifeless lump. Instead of getting the paper, she returns to the house to get me, and we walk down together. The blood in the grass is bright red, and the body is still warm. It has not been dead for very long. Only the head is missing, and there is a puncture wound in the rib cage right over the heart. Part of the neck is gone too; otherwise it is intact. We decide to salvage it for eating, and Cynthia walks back to heat some water for scalding. I check the entire fence line of the pasture looking for holes or evidence of something digging its way under, but find nothing. There is no sign of the missing head either. Whatever killed the goose, came over the fence.

As we pluck the feathers we discuss what it could have been. A dog would probably not have stopped at killing one, and a coyote or bobcat would have tried to carry everything away, unless it was scared off. A weasel apparently just kills the bird and leaves it. What about a raccoon—would it be able to sneak up on a large goose in a watchful flock, and then only take off with the head? It does not seem likely, but then none of these predators seem very likely, and the decapitation remains a mystery.

Three days later, while making my evening round to feed the animals, I notice that one of the pilgrim ganders is missing. And there, at the far end of that pasture, is another lifeless body with a missing head. It is not a recent killing; the blood is brown, and it appears a bit bloated from lying in the hot sun all day. Nothing has been eaten except the head, and there are no wounds on the body. I'll have to bury this one in the vineyard. Again, I inspect the fences, but they are all in good condition. I feel defenseless and upset. Will we lose both our flocks this way?

Then it stops for a while, before two more are killed. Somebody tells us about a similar event of chicken killings, where a decapitated hen had been found every other morning for about a week. It had been the same time of year. But before it had stopped, someone had spent a few evenings and dawns watching the flock, eventually discovering that it was a great horned owl that came and killed the birds. Thinking about this big silent bird from the night, powerful enough to kill a fifteen-pound goose whenever it wants to, just to eat the bony head and neck while leaving several pounds of good meat behind, I suddenly sense another order of the universe, a wild force following a different design and purpose. It occurs to me that what we, from our human point of view, only see as random, wasteful, or chaotic, might in fact be some important pattern beyond our small knowing.

The grapes are all dark blue now—black at a distance—sweet and juicy, but the acid still comes through as a bite on the tongue. I have started tasting them on a regular basis, moving between the west side, the upper east side, and the lower east side. It is still easy to tell that they are not ready to harvest. But even though they are not perfectly ripe, scats full of seeds and skins tell me not everyone is willing to wait. I wonder if it is a possum, and if not, a raccoon?

I know from personal experience that one can make bad wine from good grapes, but poor grapes will never make a great wine; it takes good grapes to make good wine. For the grapes to be good they must have the proper balance between sugar, pH, and acidity. The challenge with Maréchal Foch is that fairly early in the ripening process the grapes get so sweet that the sugar masks the acid, tricking the tongue into believing that ripeness has arrived. The first year we had a crop, I was fooled by this. I thought the grapes were simply delicious with no trace of acid left in them, and therefore had to be ripe. So we picked. But once the fermentation had completed the transformation of sugar into alcohol, the acid spoke up in all its tartness. That wine is still sharp on the tongue.

Having learned my lesson, I now monitor the grapes with modern equipment: a refractometer for the sugar, a pH-meter, and a burette for the acid titration. But I am also trying to memorize something which is much more elusive—the flavor of perfectly ripe grapes. I know it is there somewhere behind that cloying sweetness, a rich, full, almost spicy taste. I let the grapes melt on my tongue and try to remember the flavor from the previous harvests. This is the fifth crop and I still have not mastered this memory.

After selecting the best-looking of our three Orpington roosters, Cynthia and I slaughtered the other two because of all the fighting and crowing, and today the boys found the remaining one dead in the coop. There had been no sign of disease or sickly behavior, no warning at all. Suddenly he is down in the dust, dead and stiff. Did he find and swallow a nail? Was it a heart attack? Hernia? Too many hens for one rooster? I suppose I should have done an autopsy to try to find out, but the thought of cutting into a stiff, dusty carcass simply seemed too much, and I just buried him in the vineyard. So much for our vision of letting the flock breed and multiply on its own next spring!

Nils and Carl help out with the potato harvest. There is a certain basic digging technique that I try to convey to them, about approaching each plant carefully from the sides, and getting the fork in under the potatoes, so that they are not pierced or sliced. The soil is dry and they come out nice and clean. I weigh the yield of each row so I can calculate yield per potato planted; my plan is to track this ratio over a number of years to discover which varieties do well in this soil and climate.

Then we spread the potatoes out in the shade under a tree to sort them. The smallest get composted. All damaged potatoes are put in a bag for immediate consumption. We select nice looking ones for planting next spring and put them aside. The rest are returned to the wheelbarrow and rolled to the bins in the cellar. The total weight comes to well over a quarter of a ton. I remember my father, who always said that you plant potatoes in newly cleared ground because potatoes "break" the soil. But it is not the potatoes that break the soil; it is we who break the soil by furrowing, hilling, and turning the soil over at harvest.

One day I realize that the hummingbirds are gone, that the territorial fighting over the flowers of the butterfly bush has ceased. It was such an inconspicuous departure that I noticed nothing until the accumulation of absence became so great that it finally alerted me. Then it strikes me that the swallows have left too, slipping away while I was preoccupied with my vigilance for the grape eaters: the robins, starlings, and cedar waxwings.

At dawn, when bird calls reach into my dreams through the open window, my first thought is that the starlings have arrived. I get out of bed and start scanning the vineyard, but I cannot see any birds at all. After a while I discover that the sounds come from a single garrulous flicker. The red under his wings flashes every time he makes a move. Even though he eats grapes, he is not something I worry about—flickers do not come in flocks.

A few days later a handful of robins move into my neighbor's tall fir trees. We have heard the propane cannons going off down at the blueberry farm on the west side of the hill for almost two months, and maybe they are running out of blueberries. This could be the scouting party. The first time we had a crop coming I thought we could wait them out. "Ten birds," I said to Cynthia, "what's that? Let them eat some, there's plenty for them too!" But harvest was further away than I thought and the flock grew exponentially. Only a couple of days later there were literally hundreds of birds swooping down into the vines, calling out with excitement, while we were frantically trying to get the bird netting on to protect what was left. Ever since, we do not give them much of a chance. As soon as the grapes taste sweet and we see the first birds coming, we start hanging up the nets. A few birds will persist in spite of the nets, find gaps and openings, and eat whatever they can reach.

Green military helicopters fly overhead with increasing frequency. They seem to originate from some point northeast of us, fanning out south as they go. Some fly along the foothills, others head straight for the valley, and occasionally we can see them scan a certain spot, weaving back and forth just above the tree tops. I have never seen them land anywhere. Occasionally one of these helicopters will come right over us, flying low, and chickens, geese, and sheep scatter in a brief moment of panic. Strange rods protrude from the front or the sides. That might be their infrared detection equipment, able to spot the unique temperature of marijuana plants.

This afternoon one helicopter, returning from the valley, decides to check out our neighbor's twenty-acre clearcut. It comes down low over the brush and the brambles and makes two passes before lifting and continuing north, and I get the privilege of watching the government spend taxpayers' money at close range.

The only man-made thing that I have ever found on this land was a rusty four-and-a-half-pound iron weight with a hook on it. The thing got stuck in the tractor tiller when I was preparing a small patch of ground for planting. It is probably not particularly old. Someone said it was the counterweight on a steelyard, a scale used to weigh hay bales. As farm country this land is very young. Working here, I often sense the ghosts of the ancient trees around me, reminders of an untamed life, of other possibilities.

Every evening now the geese are fed some grain to complement the scant forage. They have come to expect this. Whenever one of us approaches their enclosure toward the end of the day they start honking and calling loudly, running up and down behind the fence as a reminder of our duty toward them. They spot the grain tub instantly and follow it around wherever it is taken; they know what it contains and will not be fooled.

When we had a smaller flock, I used to pour the grain into a small metal trough, but now it makes feeding too crowded. Instead, I just broadcast the wheat directly on the ground. Sometimes Konrad loses his wits and tries to attack the tub. Once the grain is on the ground he usually makes one or two half-hearted mock attacks before he joins the tumultuous pecking. Ruled by his hunger, he even snips at his own offspring if the competition gets too severe. While the birds eat I clean the buckets and fill them with fresh water.

I notice that one of the young birds is persistently pushed out from the circle of eating geese. Konrad pecks it hard, and the young one squeals and runs off for a few steps, then tries to re-enter the flock among its siblings, but they chase it off too. While the others eat, it meanders around aimlessly and undecided. It looks at a few spears of grass growing there but does not eat them; walks over to the bucket, starts to dip its head into the water and preen, but stops that too, apparently self-conscious with me sitting there. It is not rejected because it looks different —basically it has the same color pattern as its seven siblings. This goose is smaller than the others, but is its size the cause or the result of not being allowed to eat the grain? It seems to be personality, even though I cannot be sure. Maybe this has gone on for a long time.

Later I ask Cynthia what she thinks, and her gut feeling is the same as mine: that it is a male. When we slaughter it next month we will know.

The arrival of a yellow leaf here and there in the vineyard marks the beginning of the final phase of the growing season. It is the first and oldest leaves on each cane, the ones that spent most of the summer shaded by others, that turn first. Perhaps the yellowing is the ultimate result of the powdery mildew taking its toll on the canopy, even though the signs are not obvious; perhaps it is the inevitable result of the ground running out of moisture and the vines shutting down inefficient leaves in the last push of maturing the grapes. As the dead leaves fall to the ground, the canopy opens up and the harvest becomes visible and real for the first time. Our waiting takes on a new kind of urgency.

The weather is still dry and I decide to clean the wood stove and sweep the chimney. I find a dead starling in the stove, sooty and almost weightless, triggering memories of strange sounds in the chimney early in the summer. Afterwards, after making my way through the scuttle up to the ridge of the roof, Cynthia hands me the steel brush and each of the fiberglass rods. Cleaning the chimney offers a certain elevated perspective. From here I can see the pond across the road, groves of Douglas fir, fields of Christmas trees, even the snowy peak of Mt. Hood. Sitting on top of a roof is how I imagine a philosopher's mind: lofty and detached, balanced on the edge of vertigo.

Looking around I realize that thirty years ago there was only one house between our little ridge and the next, and that was the original Kamrath homestead—the one with the pond—and it remained the only house for sixty or seventy years. Fifteen years ago three more houses had come. Now there are five, and our house is the most recent. Beyond the hill towards Portland, the suburbs come marching our way. It makes me think of a puzzler I once read in a book on ecology: You have a pond and an algae capable of doubling itself every twenty-four hours, and you introduce the algae into the pond. It will take thirty days for the algae to cover the entire pond surface. The question is, on which day was the pond's surface only half full?

When I read this, the answer was not provided in the text; it was something you had to look up in the footnotes. There it said that the pond was only half covered on the twenty-ninth day, only quarter covered on the twenty-eighth, and so on.

During the last days of the month, early in the morning, passing through our small orchard on my way to the winery, I notice that the almond husks have started turning yellow, splitting and dropping to the ground. Later in the day I bring a large basket and the orchard ladder from the garden shed. In less than an hour I have picked everything and the basket is filled to the brim. The wind feels cold, so I find a sheltered spot and start husking. The color of the husks vary from a healthy and fleshy yellow to a shrivelled brown that looks like rotting leather. The fresh ones look so palatable in comparison that I have to taste one, but it is so bitter I have to spit it out. On some of the husks a clear sticky sap has oozed out and formed a perfect drop like a tiny crystal ball. When I am done, I have about a gallon and a half of nuts. I compost the husks and spread the nuts out on some newspapers on the hearth to dry.

Hall's hardy almond is not a true sweet almond, but a winter-hardy, early ripening peach-almond cross, in which the flesh of the fruit has been reduced to a rind, and the size of the seed increased to that of a small almond. The flavor is not perfectly sweet, but more intense, with a slight bitterness. The shells are much harder than a regular almond, but still easy enough to crack. I collect the kernels in a small brown paper bag. Out of curiosity I weigh the harvest when I am done: just about one pound. It took three to four hours for the harvest and the shelling. Had I been a hunter and gatherer, one pound of almonds, with its sixty grams of protein, would have been enough for one adult for one day.

With harvest approaching it becomes increasingly difficult to focus on other tasks. I often find myself wandering around in the vineyard, simply waiting, occasionally adjusting a net, looking at the grapes for signs of distress or disease, tasting them. One day I run into a juvenile flicker with his foot caught in the netting on top of a post. I have not heard any calls and wonder how long he has been there. As I approach he desperately calls out while trying to fly off, and I can see the thin plastic mesh cutting into his leg just above the foot. I grasp him from behind to fold his wings, then cut the netting with my pocket knife. Just as I intend to release him, he surprises me by twisting his head around and giving my hand such a good peck that I automatically let go.

In the evenings, our house on top of the hill remains in swirls of warm air, but down the driveway there is always a certain point where the air is noticeably cooler. Walking into it is like passing through an invisible wall. The cool air smells differently too. If I walk down the driveway just before dusk, I almost get to the gate before it happens. The later it gets in the evening, the further up the hill the cool air advances. But where does it come from? From water moving deep in the ground, or from the spring-fed pond across the road? Or all the way from the creek down at the bottom of the hill? It would take some hiking around in the dark to find out.

Ten

We left Sweden again after three years, and that was twelve years ago. It is hard to point to any one particular event and say: this is when I made up my mind to stay in the U.S.; or this is when I decided not to go back. The years have merged with each other, and now it is difficult to separate one from the other. Was it the scholarship that brought me back to Oregon to complete my academic studies? Or that this scholarship was extended, so that instead of having to return, I could stay on for another year? Did it have to do with the fact that my degree meant practically nothing in terms of employment? That there was no job in Sweden to go back to? That I began having doubts about working in academia anyway? Or that translation work began appearing at my door, and that it was a job that could be done practically anywhere?

Or wasn't it work at all, but more related to the reality of being a family with two infant boys, and that moving back to Europe again suddenly seemed overwhelming and perhaps the wrong thing to do? That having children made us want a home in a way we had never had one before? Or was it simply that the appeal of the country had resurfaced again after several years of living in that strange compromise known as the suburb, neither city nor country? That my mother had unexpectedly died, and that Cynthia's parents were in poor health? That I no longer felt particularly homesick for Sweden since some part of me had redefined what I meant by feeling at home, and that it was this new emerging understanding that was the reason behind our decision to look for land?

If you ask me if I believe in love at first sight, I will say that I do, and quickly add that I don't think it is limited to a feeling between two people. I know there was love at first sight when we first walked across these eight acres. But what I understand now is how much it takes before one is able to fall in love.

In the winery, vats and barrels have been washed, rinsed, and carried into place. The machine that crushes the grapes and removes the stems, a contraption known as a crusher-destemmer, has been cleaned, greased, and tested. The picking shears have been counted and inspected. A small first-aid kit with band-aids and wasp-sting relief has been assembled. Almost two hundred buckets have been brought out from storage and rinsed off. Pickers and customers have been alerted that harvest is approaching.

For the grapes, each sunny day will help reduce some of the acid and add another fraction of sugar, so we keep waiting. But since mornings are cool and foggy, with the sun not breaking through until late morning or noon, ripening moves at a snail's pace. It is not all bad. One theory holds that the longer the grapes must hang on the vine to achieve maturity, the more complex the flavor in the wine. This is commonly referred to as "hang-time," and our cool site is certainly giving us that. The main question is, how long is it safe to wait? Since we rely on friends for pickers, it will take us three or four days to bring in all the grapes. Several times a day we listen to the weather radio to see where the big Pacific storms are and what they are doing —remaining at sea or heading for northern Oregon? The chance is, of course, that once it starts raining seriously now, it may never really stop again until spring, and cold rain could ruin the grapes quickly.

Three Steller's jays keep raiding the mature sunflowers in the garden. They love these seeds. They can even eat them sitting upside down. I have often wondered how these shy forest birds, which so rarely fly into the open of our yard when we are around, know about sunflowers. No wild sunflowers grow in this forest, so how do they know that these big, brown, round things on top of tall sturdy stems contain delicious seeds? How do they recognize it—is it a smell?

While contemplating this small mystery, I sift the last bin of finished compost for the season. It is the ninth bin, and when it is done I will have made approximately ninety wheelbarrow loads for the garden since spring, perhaps adding an inch of sifted compost throughout. And in spite of all these applications, the stuff just seems to disappear, leaving only pieces of eggshells and black grape seeds behind. It is a process that takes time; I have read studies that claim that it takes about ten years to build up a decent humus level in the soil when an inch of compost is added annually.

Kindling a fire in the early evening, burning small branches gathered on these eight acres to get a good bed of coals, I feel that the land accepts me and that I belong to it, in spite of not being native, and in spite of all the machinery and technology that stands between us. The connection between the ground and my feet feels alive and magnetic. I never knew I would come to live here, and what I do now is not what I had prepared myself for. Still, work on the farm comes naturally, as if following a path of least resistance. As soon as an idea is realized, it becomes clear what must follow, and it has been this way since I came here. Not that the past is always easy to abandon, or that what one finds here is exactly what one wants, but the place suggests that possibility of what could be. Ten years ago it was eight acres of pasture. Now there is a house, a vineyard, farm buildings, animals, and it is still growing and changing. It is the old story of the New World, that powerful dream of being able to reinvent oneself into something new, of transformation and change.

    I brush four slabs of venison ribs with a mustard, garlic, and olive oil mix, and keep turning them over the coals as dusk settles. The colors of the sky are fading in the west and the dark green firs turn black. The silhouette of a small owl, the first I have seen in a long time, glides in noiselessly from the neighbor's trees, circles the yard, and disappears out over the vineyard. Cynthia brings fresh vegetables from the garden, a large bowl of cherry tomatoes, a plate of sliced cucumbers, bell peppers, and carrots. We drink from a bottle of St. Croix, a wine with a taste of wild berries.

The summer has not been exceptionally hot, but there has not been much rain either, and the grape clusters are beginning to show it. It is most apparent along the top of the ridge, where one can see the first signs of grapes turning to raisins. In the lower parts of the vineyard there must be moisture left in the soil, because there the grapes remain plump and juicy. They taste deliciously sweet and spicy. Testing the vineyard samples reveals that the Maréchal Foch now have an average total acidity of eight and one-half grams per liter, a pH of 3.2, and a sugar level of almost twenty-five degrees Brix. In short, they are perfectly ripe.

Once the decision to launch the harvest is made, our lives accelerate into a vortex of grapes. I sit down at the phone to organize the pickers and firm up dates and times with this year's customers. In the vineyard, Cynthia starts unhooking the nets and shoving them into stuff sacks. Once I am done calling, I go outside to join her. It is slow and repetitious work and we keep at it until dark.

As long as Nils and Carl are willing to help with the harvest, we let them stay home from school. As this is the culmination of our agricultural cycle, it seems only natural that they should be invited to participate. They get the same pay as a commercial picker, a dollar for every twenty-pound bucket, and the harvest is a good opportunity for them to earn some money. Over the years they have become increasingly skilled at picking, to the point that we now actually rely on them for help.

During the first day of the harvest the conversations among the leaves are inspired and amiable. The weather is just right, neither too cold nor too hot. There is still a sense of novelty to the dark blue clusters filling the white plastic buckets. "I like the clicking sounds of all the picking shears," Nils says with a big smile, happy to be part of the busiest days of the year. I am too busy to pick. I am either helping Cynthia remove more nets or lugging full buckets out of the rows to the scale on the tractor trailer, bringing back empty ones. I also track the total amount of grapes coming out of each five-row section. Once a section is picked, I compare it to the same section in previous years and quickly get an idea of what the total harvest might bring. The news is good: yield is up from last year by about ten percent, so maybe there will be five tons this year. When the trailer fills up with buckets, I haul them from the vineyard to a place in the shade where the buyer can pick them up. Unlike most home winemakers, who want us to process their grapes, this man will take them as is. It makes for an easy start of the harvest.

The crew is good and the picking is done several hours before the grapes are to be picked up. The fog burned off midmorning, and the day has turned pleasantly warm. Now there are hats, scarves, coats, and sweaters scattered like scarecrows in the vineyard. Everyone sits down with purple hands to eat and drink and celebrate the new vintage. As the meal comes to an end and the conversation slows down, my thoughts return to the vineyard. There are nets to remove for tomorrow, and dark comes early. Soon, Cynthia and I are back among the leaves in another section, unhooking nets and stuffing them into sacks.

I return to the vineyard in the evening, and as I walk through a few rows in the darkness I hear nets rub against the leaves in the light breeze. It is a dry, rustling sound, reminding me of how worn and leathery the leaves now are. They have done their utmost to ripen the crop and they are dying. I doubt they will hang on much beyond the harvest. As I walk between the long nets they seem like sails, and I get the sensation that the whole vineyard is a ship flying through the night. Behind me, I am leaving a sky illuminated by the lights of Portland, sailing south toward the black silhouette of the mountains. The ridge line is smooth past Goat Mountain, then turns into a series of humps and crinkles, then a final elegant rise ending in Seosap Peak with its blinking tower beacon. Scattered high on the mountain side I notice three bright lights that I don't remember from last year, and I am surprised to see houses that high up. Then I let my eyes climb higher as the vineyard sails out among the stars.

During harvest, there are always a number of bird nests discovered in the canopy, mostly empty goldfinch nests, small padded bowls neatly woven out of wool and grass. Occasionally a deteriorating robin's nest—deserted a long time ago—is found. This year someone finds an abandoned goldfinch nest with five white eggs. Several harvests ago, when the last grapes were picked in the rain, we found a couple of green tree frogs clinging to the canes, blissfully welcoming the drops. Last year we found a grey wasp nest the size of a large grapefruit built around a cane, and, after we made sure it was empty, we cut it loose for the boys to take to school. That same year, while putting up the nets, we also ran into several underground wasp nests established in empty gopher holes. We managed not to get stung, but picking around them was not much fun.

The last grapes are for ourselves, for our own wine. It is the fourth day of the harvest and Cynthia and I are getting thoroughly tired of it. The magic has faded, and picking has become nothing but the repetitious, mechanical movements of tired hands and arms. My back is sore from all the lifting and carrying and I have seen so many clusters that the image appears even after I close my eyes at bedtime. Both boys gave up yesterday—they had earned enough and were ready to go back to school. In the vineyard, the conversations get increasingly silly. Someone tired of standing starts a long yarn about what they want for next year: a motorized picking chair with a bucket holder, sort of like a cross between an all-terrain vehicle and a modified electric wheelchair operated with a joystick.

With all the nets down I am finally picking too, daydreaming of Walt Whitman and his descriptions of the singing that went on while people worked together. Before the machines came, people everywhere sang while working. Different tasks had different songs, helping people keep a certain rhythm while taking their minds off the drudgery. I am thinking that maybe I should hire a singer for that final day of picking. But what kind of songs? Folk songs or Italian opera or . . . ? I never finish my thought, because out of nowhere, a minivan pulls up in the driveway and a group of smiling Japanese tourists emerge. Everyone welcomes the interruption and some of the pickers pose for photographs, waving shears and passing out clusters of grapes. I smile at the irony of the situation, that after years of traveling abroad watching others live off the land, I am now the one visited by others. I take it as an indisputable proof that we have at last settled into an authentic way of life ourselves.

The weather is still holding as we finish the last of the harvest, but just barely. As the pickers leave, I haul the last buckets to our small winery and put them under cover, then head up to the house for a much-needed break. It smells like rain, and the wind gusts are getting cold. The year it started raining at the end of the harvest, the temperature plummeted and the rain did not let up for two weeks. Those who waited, thinking it was a small front that would quickly blow over, had to pick in ice-cold pouring rain.

It is dark and raining outside when we finish dinner, but there are lights under the roofed area where the grapes are crushed. Before Cynthia and I start, we put about fifteen percent whole clusters in the bottom of the vats. The idea behind it is twofold. First of all it leads to a special kind of fermentation inside the uncrushed grapes, a process known as carbonic maceration, which yields a uniquely fruity wine. Secondly, since the sugar inside the whole berries is less readily available to the yeast multiplying in the must, this technique prolongs the fermentation time, which results in increased extraction of color and flavor from the grape skins. The combined effect is a wine that is both fruitier and richer.

With the whole clusters in place, I begin dumping buckets of grapes into the hopper of the roaring machine. A rotating auger conveys the clusters to a pair of cylinders that pull down the grapes and crush them. Below, there is a quickly spinning shaft with a double helix of metal spikes that kick out the stems at one end, while the juice, skins, and seeds are allowed to fall through a screen. A catch basin funnels the juice and pulp into a bucket, which Cynthia then empties into the fermentation vats. It is an ingenious Italian contraption that allows two adults to process a ton of grapes in just about an hour.

We walk back to the house in the rain, tired but content. The crop has been safely brought in, and there was more of it than we thought: 12,368 pounds, which should make approximately 4,123 bottles of wine. There are still buckets to wash and things to put away, and the picking shears will need cleaning and greasing, but the rush is over. There are no more customers to call, no more pickers to alert, no more birds to worry about, no more nets to stuff, no more storms to watch out for.

After the first heavy rains the Oregon juncos return. Several weeks ago, looking in vain for chanterelles up in the mountains, I had seen lots of juncos there and none at home. Now they have come down from the higher elevations and eat the tiny pigweed seeds with a hurry and hunger that surprises me. They move around in small flocks and are easily startled. One day I find an Oregon junco sitting out of the wind under the stairs. I bend down to watch him and he doesn't move. He looks back at me with tiny obsidian eyes expressing nothing. He has probably flown into a window. Later I find some brown and black tail feathers in a tuft of grass around the corner and I suspect that one of the cats found him too.

The unhurried pace of winemaking is a pleasant counterpoint to the rush of the harvest. Because the grapes were cool and cold weather followed with the rain, the fermentation is slow to begin. But during the evening of the second day I hear the must bubble and crackle, and know that the transformation into wine has commenced. Billions of bubbles of carbon dioxide soon rise in the must, lifting the skins up to the surface, forming a stiff cap. Since much of the color and flavor in a red wine comes from the skins, one must break the cap several times a day and stir it back down into the must. Three times a day I walk down to the winery to stir the vats, inhaling the fumes and contemplating the vintage. I usually taste it daily, not to see what the vintage will be like since it is much too early to tell, but to see how much sugar is still left and how many days the process might last. Normally I leave it on the skins for a week or two.

As the fermentation accelerates in the vats, the temperature increases, until the third or fourth day, when it peaks and slowly starts to decline. At its peak, the temperature often reaches into the nineties even if the ambient temperature is cool. During this whole process, enormous quantities of carbon dioxide are generated, and for the first week I have to open the door for several minutes before I can enter. Still, there is so much carbon dioxide whirling around that it quickly generates a headache. But the lack of oxygen does not seem to bother the small, reddish fruit flies that have somehow managed to multiply and survive. Some years seem to be worse than others. The fruit fly is supposedly a carrier of the vinegar bacteria, and could theoretically, if present in sufficient quantities, inoculate the fermenting must and convert it into vinegar. If the clouds get too big I sometimes take a vacuum cleaner and eliminate as many as I can.

I wake up to a day of clear skies and notice that there is snow on the mountains again. I walk with the boys down to the bus stop at the corner of the empty vineyard. The air is crisp and I enjoy the feeling of having no obligation to work there. Most of the leaves have disappeared, but those that still hang on have turned a clear yellow. While we stand around waiting for the school bus we notice the hump of a dead animal down the road. We walk over and discover that it is a ringtail raccoon. His intelligent bear face stares into the distant sky and his dexterous hands clasp nothing. Living fairly far from a creek, I never thought raccoons moved through here.

Getting the mail in the afternoon, I remember the raccoon and walk back to take a better look. Someone has stopped and cut off his tail, leaving the body right in the middle of the road. I would like to skin it and tan the beautiful pelt but do not know how, so I pick it up and throw it in among the Christmas trees on the other side of the road.

Pressing the wine brings the harvest to its final conclusion. Cynthia and I bucket the raw wine out of the vats and pour it into the wine press, collecting the free flow in pails. This is poured through a strainer into the stainless steel barrels. The wine tastes so raw, harsh, edgy, yeasty, and effervescent that it is difficult to imagine it as a finished wine. I still would not drink this for dinner.

Once the basket is full of skins and seeds we press it lightly, letting the wine take its time to trickle out. After pressing we scoop out the press cake with our hands, which turn a deep inky blue, almost black. Lemon juice will remove most of the stains, but the edges under our fingernails and the cracks in our hands will remain black for days. The press is filled three times before we are done. The wine will now be left to settle for a few weeks in the stainless steel barrels before it is siphoned into the oak barrels in the cellar below.

For several days dawn is exchanged for a skim-milk fog. Fall is here and the smell of wood smoke mingles with the vapor. Every blade of grass has turned silver from the tiny drops. All day the fog expands and contracts like some strange tidal mist, moving sideways through the vines and trees like waves, wetting everything. The fog condenses on the roofs and trickles down the drain pipe, making diminutive sounds. It condenses underneath roofs and overhangs, even on walls. Luminous patches of light come and go in the heavens as the sun follows its arch. At night, when quiet has come again, it smells like earth and potatoes, and I can hear the fog in the trees, big drops falling and making loud noises as they hit the wet leaves.

Eleven

As a permanent resident alien I have been granted permission to live and work here. I pay taxes and I would have to serve in the armed forces if the U.S. government so required. I can own property. I don't have the right to vote or to sit on a jury, and in a sense, it is taxation without representation. On the other hand, even though I don't live in Sweden, I can still vote there through the local consulate. It is a fair arrangement, and an old one dating back to the laws regulating the free citizens of the Greek city-states of 300 B.C. Sometimes it is a source of irritation not to be able to cast my ballot locally, but it is part of the price I have to pay for maintaining another allegiance.

Discussing this, I am occasionally asked why I don't become an American citizen, since it would eliminate that final barrier to full participation and involvement. Why don't I? Unlike many other countries, Sweden would not allow me dual citizenship, so if I were to become an American citizen, I would have to renounce my Swedish passport, and that would lead to my children losing their dual Swedish-American citizenships. Some part of me is unwilling to do that, because I feel that I owe them that link to their family roots. Besides, I still feel a personal loyalty to the language, to my own family, childhood, and youth, to certain aspects of Swedish history and customs, and to certain culturally inherited values and beliefs, and all these things have become so inextricably intertwined with my own sense of who I am that I have no desire to renounce them.

And yet I love this land.

After the first heavy frost, the Japanese persimmon tree drops all its leaves in a perfect circle right below it, leaving a lace of black branches against the sky. The stiff, fat leaves have turned red with streaks of green and yellow, and the pile looks like money, like bills of some cheerful foreign currency. Yes, I am ready to exchange the dull dollars in my wallet for these leaves. I am ready to follow.

A stretch of clear weather settles in and I start replacing the old, broken field fence along the north side of our property. The south and the west sides were done last year, and I should be able to complete what is left this winter and be safe from roaming dogs and coyotes. I untangle salal, blackberry, and cascara buckthorn saplings that have grown into the wires, then attack the roots with a mattock. Eventually I am able to remove the broken fence from the rusty T-posts, shake it clean and bundle it up. Perhaps it can still be recycled as scrap metal. Under some leaves next to one of the old cedar posts I startle a mottled, black, ten-inch salamander. It is the first one of this kind that I have ever seen. I am surprised at how rapidly it moves. Not at all like the small, brown, Pacific newts with their beautiful orange bellies and slow, dream-like movements.

I pull the old T-posts out of the ground and replace them with eight-footers. This will accommodate a higher field fence and two strands of barbed wire on top, and it will be practically deer-proof. I replace the occasional wooden post with a taller one too. I dig the holes by hand, recalling that cowboy poet who, when asked about the Ph.D. after his name, claimed that he had earned it in post-hole digging.

Later, I light a fire and start cutting the brush. Everything with a diameter larger than one inch I save for kindling, the rest I throw on the flames. I have not done this kind of work for a long time and my cuts lack precision. As I clean a long forked filbert trunk, I have to force one branch down with my foot to be able to cut the thin top off, but as the axe severs it, it kicks back with tremendous force and whacks the cheek bone under my right eye. The blow is so fierce I let out a long, loud yell. I return to the house and apply ice, but still I get a great big bump and the first black eye of my life.

In the barrels of new wine, yeast cells and sediment riding the minute bubbles of rising carbon dioxide have come to rest in the lees, and because of this, the wine has cleared considerably. The yeasty flavor is almost gone. Unclouded and fresh, the wine is still more purple than red, with a slight effervescence tickling the tongue. The fresh taste of berries is certainly there.

At this point Cynthia and I empty the oak barrels of last year's wine, rinse out sediment and purple tartrate crystals, and get them ready for the new vintage. We siphon the wine into five-gallon glass carboys, carefully labeling them with the name of the barrel. The American oak barrel has only held wine for one year, and has imparted rich vanilla and cinnamon flavors and an attractive, voluptuous smoothness to the wine. The second barrel—made of Slovenian oak—which has held wine for three years now, has made it slightly stern and tannic, adding a bit of structure. The last one is an old French barrel, and its contribution consists of a more subtle and neutral enrichment and concentration of the wine. The final mix will be a gathering of all these aspects.

Several years ago I made two barrels of pinot noir, with grapes from one vineyard picked on the same day, and vinified in absolutely identical ways. The sole difference was that since I just had one oak barrel, half the wine had to remain in stainless steel. Twelve months later it was like night and day; one could not believe that it had been the same wine. The elixir coming out of the oak barrel was deep red with a purple tinge, smelling fruity and spicy. The wine from the stainless steel barrel was pale and unattractive, as if it had been nearly starved to death inside its metal prison. It was clear to us that red wine needs to react with oak, extract flavor from it, slowly evaporate through the porous staves and go through its process of concentration. Ever since, I have made sure that I have had oak barrels for our red wine.

The tenth of November in Sweden is known as *Mårten Gås*. The words simply mean Martin Goose. At the time of the early crusades a third-century Frenchman, St. Martin of Tours, apparently became part of southern Scandinavian culture, associated with the eating of goose. Today no one knows what the real reasons were, if St. Martin liked goose or if the church simply needed a saint for recently converted heathens. Possibly it was a much older tradition for which St. Martin became the officially approved mask. In a farm culture where geese were used to glean the grain fields after harvest, the tenth of November was a good time to get rid of surplus birds before the onset of winter.

It is time to reduce our flocks too. Our first geese, a mongrel breed showing signs of the blood lines of Toulouse, brown African, and Embden, will be replaced by the relatively quiet and gentle Pilgrim geese. The old flock has grown strong. Konrad, the alpha male, has developed quite a militant nature. He is clearly in charge. He does not like anyone changing the water and will come charging when someone enters. He does not like the grain being brought every day, and will usually attack the bucket from which it is poured. The boys refuse to go in to the pasture; Cynthia never enters without a stick. I never turn my back to Konrad for even a second. More than once he has beaten me with his wings and bitten my pants, and more than once I have had to grab him by the neck, swing him around and throw him back into the hissing and cheering flock. The plan is to eliminate Konrad first, followed by the second in command, Uncle Ernie. Then eliminate the rest of the flock a few at a time.

After lunch Cynthia and I get a cage ready and walk over to the pasture. It is as if Konrad senses that something is up, and he keeps the flock at a safe distance from the fence while maintaining a keen eye on us. I start to change the water to see if he will come charging. But today he stays back. On our side of the fence I start hitting the handle against the empty plastic bucket. Somehow this must remind Konrad of fighting buckets, and he forgets his premonitions and immediately comes charging. But as soon as he sticks his head through the field fence to bite the bucket or us, I grab his head, pulling his neck until his body is tight against the wires. Then I ask Cynthia to hold his head while I jump the fence to fold his beating wings. I tuck his body under my left arm, secure his head with my right hand, and take him to the cage.

All afternoon he honks loudly to the rest of the flock, and the flock answers. Sometimes they call him and he replies. This continues into the night. The trumpeting is very loud and I wonder what the neighbors think, whether they like the sound or find it annoying.

In the morning I take him to the chopping block out by the wood pile, where I have hammered in two large nails about an inch apart. The neck barely slides between the nail heads and I pull his head back tight against them by pulling his body backwards. Konrad is absolutely still. I hold him like this while saying a brief thanksgiving to him. Then I apologize for taking his life and bring the axe down right behind the nails. A second blow is needed to completely sever the head, which bounces off into the grass. The one eye facing me remains open and clear. It looks surprised, as if this was impossible. It is impossible and possible at the same time; death being as incomprehensible as birth. The bright red blood is pumped out over the chopping block, the grass, my rubber boots, and my jeans. I grab his feet and, with some effort, because he is heavy, hold him at arms length. His big wings start flapping heavily and regularly, like a swan or a heron lifting. He is flying now like his ancestors. He is graceful in his dying and I admire it.

After scalding we pick the feathers first, then the down underneath, which Cynthia saves for something in the future. Perhaps a down comforter for our bed or vests for the boys. Cleaned he weighs thirteen pounds. I will soak him in brine for two days in the refrigerator, Danish style, then boil him for two or three hours in the company of fifteen white peppercorns.

Throughout the day the flock keeps calling for Konrad.

The inevitable fall rains settle in and the sun disappears behind a canvas sky where the clouds practice endless variations of grey rain. The days are short and the temperatures are still fairly mild. Dressed in full rain gear and slipping around in the mud, I spend a few days clearing the last of the brush along the fence line. There is other work to do outside, but nothing that can't wait, so I spend more and more time inside the house.

And there, back in my reading chair, I finally stumble across the answer to my question of why the French-American grape hybrids never received widespread acknowledgment. I certainly knew René's answer was a red herring, and even though I had my suspicions, I had never heard anyone articulate any convincing arguments. At the end of *The Great Wine Blight*, a fascinating history of the phylloxera devastation of the European vineyards toward the end of the last century, George Ordish lines up three main reasons: the power of the established French *chateaux*, fearing the competition of the productive hybrids; the French government worrying about a yet bigger wine surplus and all that it would entail in agricultural subsidies and support; and the chemical trade, standing to lose the highly profitable business of selling chemicals to fight the diseases the hybrids are resistant to. In retrospect, I realize that the question I should have asked René was, who paid his salary during his years in Brussels?

Returning from a walk, Cynthia and I find that someone has dumped a doe's head in a cardboard box just past the sharp curve along the road. It is still buck hunting season, so someone either poached it or made a fatal mistake. A few days later the head disappears, but then a deer carcass appears instead, very close by. It is simply thrown in the ditch along the road together with the other garbage there, the water heater, the worn tires, and the old mattress. A few days after that the carcass disappears too, and since the other garbage does not disappear, I wonder if it could be the cougar that has been spotted several times by the new golf course, just a few miles from here.

We run into our neighbor who tells us his theory about these country-road garbage dumpers. He thinks that two people do it at night using a pickup truck, with one guy in the back throwing stuff into the ditch as they drive along. That is why the garbage is scattered at regular intervals along the straight stretches of the road. Going through the curves the fellow in the back has to hold on, and that is why there is rarely any garbage in the curves themselves, only before or after.

A former colleague calls me to ask whether I am interested in joining him and a team of technical translators—they are finally going to band together and form their own company, and would really like to include me in their group. They have a big job lined up and it will keep everyone busy for a couple of months, and they've got almost all the languages covered. They have located an office space and are working on the equipment now. Of course, he can't guarantee full-time work after the initial project, but they've found a really good marketing manager and he is very optimistic about the future. His enthusiasm makes it sound exciting—becoming part owner of a new business and having more control over one's work. In addition, there will be the opportunity to travel back to Sweden to attend trade shows and conferences to keep *à jour* with the rapidly evolving terminology. And, he reassures me, "it is still a well-paid job. You know, there aren't that many Swedes around who translate technical texts."

"What about working at home?" I ask him. I did not move to the country just to commute back into the city.

"Well, certainly not every day," he says, "but maybe part of the time. It will depend on the project, you know. You'll have to be here for some things."

I remember—all the long hours comparing pages of translated text with the original, strategy meetings, the seemingly endless discussions of this and that. Most things eventually seemed to require one's presence. "When do I need to make up my mind?" I ask.

"The sooner the better, because if you say no we have to find somebody else."

"I see. Well, I'll give you a call in a couple of days, how about that?"

"Okee-dokee," he drawls, "I'm here all week, Saturday too."

After hanging up I try to remember what it was that made me quit that job. Was it the time spent in traffic on the freeways? The stress resulting from changing texts, fickle customers, and endless deadlines, much too often conflicting with priorities on the farm? Or that I ultimately found instructions, technical descriptions, and specifications boring beyond redemption? Would I want to return to all that even if it might pay well? At what point does one feel that one makes enough money? And if I disappeared to the city every morning—who would keep up with things at home? Cynthia wouldn't have time to do it all by herself. What would that lead to—hiring a farmhand? Employing someone to live my life for me? If I said yes, wouldn't I just postpone my life here again, move away from the center and back to the periphery?

The stainless steel barrels, which held the new vintage until it was moved into the vacated oak barrels, have been cleaned to receive wine again. Into them Cynthia and I carefully siphon the content of the carboys, one-third each of last year's wine from the American, Slovenian, and French oak, making sure that we leave all the sediment in the carboys behind. This procedure makes the vintage uniform.

After mixing, the wine is allowed to rest for a few weeks, giving it a chance to combine flavors before it is bottled. Leaving it for a while also gives any trace of sediment yet one more chance to settle out and increase the brilliance of the bottled wine without having to filter it. We taste it to get a sense of what the mix is like, even though the wine is too cold to reveal its full flavor. It seems more velvety, more sensuous compared to the previous vintage, and we bring some of it up for dinner to get acquainted. The wine is still withdrawn and rough from all the handling, but we like it, and we know that it will change and that our impression of it will change too; our journey together has only just begun.

The Christmas tree harvest across the road begins with chain saws starting up at dawn. Last week someone went through and marked all the mature trees with blue flagging tape, so we knew it was coming. Only the small trees are left standing. The view reminds me of fallen soldiers after the first wave of attack.

A few days later Hispanic work crews arrive to pile the trees, and weather permitting, a small helicopter will fly from dawn to dusk picking the piles up and dropping them at the loading site. For a few days it is like living next to an airport. At the loading site they work late into the night with diesel generators and halogen lights even if the rain is pouring down, baling the trees and loading them with sloping conveyor belts into big semis. After a week of intense activity the whole operation moves away, abruptly leaving the field black and silent again.

A car load of Jehovah's Witnesses unexpectedly arrives one Saturday afternoon, even though the front gate is closed. Cynthia and I are in the middle of preparing an area for blueberry bushes. There are two middle-aged women and two young kids in the large, American-size sedan. Several groups have visited us in the past, but we have never seen these people before. Only the woman in the passenger seat leaves the car and comes over to where we are working; the others observe us from the windows. She is holding a copy of the *Watchtower* and a Bible. On the cover of the magazine I glimpse a picture of what I guess is paradise, a smiling lion cuddling a blissful lamb. I cannot help but wonder how the lion likes being a vegetarian. Sometimes during visits like these when I have been alone, I have spoken Swedish and waved my arms and pretended to know little English, but today the surprise caught us off guard.

"Hello," the woman says somewhat apprehensively. "I have come to talk about eternal life in Jesus Christ." Passion rises in her voice as she speaks. "I don't have time to talk about eternity," Cynthia says, cutting her short like she does a sales person on the phone. "Please, just leave and close the gate as you go." Sensing our stonewalling, the visitor decides to play her last card quickly. "Can I leave this free magazine with you, for later, when you have more time?" she asks, pointing with the magazine towards a pile of lumber next to the chicken coop. Cynthia looks her straight in the eye and says, "I would call that littering." The woman gives us a surprised look, and says, "At least you're keeping busy!" before scurrying back to her car. They disappear down the driveway, staring at us through the windows.

On Thanksgiving I make a small round on the farm before leaving for dinner at a neighbor's house: Thank you vines for the grapes. Thank you garden, thank you compost, thank you orchard. Thank you chickens for eggs and meat. Thank you geese for mowing and for entertaining us. Thank you sheep in the freezer. Thank you earth and sky. Thank you wife and children. The family is in good health and life is sweet to us. We live in a privileged time. Thank you for today. Thank you.

The leaves are all gone in the vineyard now, and the only thing that remains is a tangle of canes. The canes of each grape variety have a unique, identifying nuance of color, and Maréchal Foch has a slight purple tinge. It is especially prominent when looked at from a distance.

It is not raining and I walk through the vineyard just to stretch my legs and get a bit of fresh air. I stop in front of a few vines here and there. Growth in the vineyard has been good, and the canes are thick and sturdy. The trunks have firmed up and are already getting covered with several different kinds of lichen. Still, these are young vines, just about ten years old, and not even fully mature yet. How large will I ever see them—as thick as my leg? Or won't they be here when I leave this earth?

The temperature in the wine cellar has steadily dropped and all the signs of active fermentation in the barrels have ceased. There are not even any gas bubbles inside the clear plastic fermentation locks in the rubber stoppers sealing the bung holes, and there will not be any gas bubbles until spring, when warm temperatures will activate the malo-lactic fermentation again. It is the first time I can get a sense of the new vintage. I have kept the wine from the grapes grown on the east and west sides of the ridge separate, and as I sample the different barrels I can, for the first time, taste a slight difference between them. It is not much, just a more pronounced fruitiness from the west-side grapes. Overall, the wine is incredibly rich and dark. Flavors of marionberry and cherry hit the palate. There is a slight bitterness in the finish, one which I don't remember tasting in last year's vintage. I wonder if it will develop into that bittersweet chocolate flavor, or if it will disappear. But a month after its making, I realize that I am just speculating.

Wood barrels are porous, and wine evaporates through them; how much depends on the surrounding temperature and humidity. On the average, each barrel loses about a bottle per month, and this must be replenished to prevent oxidation and bacterial growth. After tasting, I uncork a few bottles of last year's wine to top off the barrels, replacing what I have tasted and what has evaporated through the wood. Before leaving I carefully wipe off and sterilize the area around the bung holes, and this is how the cellar will be tended for the next twelve months.

On the last day of the month, after almost eight years of recording temperatures, I observe that November is still the most consistent of all the months when it comes to temperature: no strange fluctuations between day and night or from one day to another, but a loyal, gentle curve snaking along the line of the old statistical average. But it is always wet, with as much as a foot of rain possible.

The wind is steady out of the east, dry and cold. Yet high above, from the completely opposite direction, rain clouds move in from the southwest.

Twelve

When I return to Sweden I feel like a tourist. Everyone is busy and I have a lot of time on my hands. Like any visitor, I can't completely grasp the value of the currency and get used to the prices of things. I can't remember all the names of the streets I used to know, and there are new buildings and businesses where the old shops used to be. Some places are so totally altered that I don't even recognize them. I notice that I dress differently, that people on the street no longer look as familiar as they used to. At first, even the lilt in their speech sounds slightly odd, as if they are speaking rhetorically as they sometimes do in the theater, or are using expressions from some other period in history. And when I speak, every now and then groping for a distant word or laboring to find the correct syntax, my friends say they can hear an American accent in my voice. I have never heard it.

If you travel somewhere as a tourist, and are received as one, it establishes a frame of reference for the conversation. The locals know you are a stranger, and will talk to you accordingly. Certain topics will never be brought up since you lack the knowledge to discuss them. Returning to your homeland is different—you are still treated as a local and are expected to know. But what is not obvious is that, even when you have tried to stay informed for as long as you have been away, you have inevitably absorbed something of the other country's perspective. After twelve years, we talk of different worlds. It is perhaps not so much a question of defining what is good or bad with this and that, but the perspective shifts everything sideways. What was taken for granted no longer seems so certain.

Walking the streets of Stockholm I discover that I am homesick for my Oregon family farm.

The old man did not say anything and did not seem to pay much attention. The young one, who was unshaven, overweight, and dressed like a logger with suspenders and heavy boots, did all the talking and introduced the old man as his father. He said they lived down the road and had seen us start planting the vineyard years ago. Now that there was a sign—a crop sign I had put up along the road in the fall saying "Wine Grapes"—he thought we had a winery and were finally open for business, so he drove up. He was surprised when he found out that there was no winery, but he still had a few questions about grapes and winemaking that he wanted to ask, and the two of us ended up talking about it for quite some time. Even though he did not fit my notion of a winemaker, it turned out that he made sweet berry wines and wanted to try one from grapes. Now, would these grapes make a sweet wine? I explained that I did not make sweet wines myself, but yes, it could be made in a sweet style.

The old man remained silent throughout the whole conversation; he did not even comment on the wine I let them taste from the barrel in the cellar. What stuck in my mind, though, is what the old man said in a surprising burst of eloquence just as they were leaving:

"That snow on the mountains, you know, is more real than paper money. Paper money isn't really real. Money in the bank means nothing, after 1929. . . ." Here the son cut him short.

"Come on Dad, we're already late, and we've kept this fella from workin' long enough." He looked at me apologetically while maneuvering his father into the car. "Thanks again for all the information. See you at next year's harvest then!"

He waved as they drove off, but the old man already seemed oblivious, his eyes fixed on the snow on the mountains.

One of the scrub jays has discovered the small, moldy chestnuts in the compost bin. It is a wary bird, nervous almost. The compost bins are about a hundred feet from the edge of the woods where he lives, and that seems to be his limit. I have never seen him fly any further into the open. He lands on top of one of the corner posts, looks around several times before making that perilous descent into the bin itself. Then he quickly emerges with an amazing load in his beak and flies back among the trees. All morning he airlifts chestnuts to his secret hiding places in the woods.

After bottling the Maréchal Foch we have a sizable private wine cellar again. Most of it will be drunk as our daily table wine, but some will be saved for the future, for the time when we will have promised buyers more grapes than is actually there, or if flukes in the weather ever get bad enough to ruin the entire crop. This time I also filled three jeroboams, great five-liter bottles that are now stashed in the far corner of the cellar, collecting dust while waiting for the right celebration.

As I stack the cases in the basement cellar, I am reminded of Thoreau, who used to claim that he had the largest private library in Concord, a minor detail being that seven hundred of the volumes happened to be the unsold copies of one of his own books. Here one of the advantages of wine over books become apparent: there is a great deal more enjoyment in drinking seven hundred bottles of the same wine than it would be to read the same book seven hundred times.

A bone-chilling fog, thick as pea soup, settles in and lingers for over a week, an unusual weather pattern on our ridge. Usually the fog stays low, along rivers and creeks. The fog is so dense that it conceals the road, which is less than five hundred feet away. Sounds are muffled and all our neighbors are invisible. Our house sits like a secluded arctic island. Cynthia finds it claustrophobic after a few days, but I like having the world enveloped by whiteness; it is like inhabiting a dream without edges. During daylight hours, the temperature hovers just above freezing, then drops a few degrees below at night, coating everything with a solid layer of frost. The wood stove is kept going throughout the day. At midday one senses the sun working on the fog from above, generating whirling currents of vapor, but the sun is too weak to burn it off. Late in the afternoon the fog stops moving as darkness falls.

A small falcon hunts the vineyard floor while the fog lasts. He perches on the line posts, and as soon as I get too close he flies a short distance and settles on a new post, continuing his watch for mice and voles. The crows appear disoriented and are much less conspicuous. They fly low and keep quiet, not at all their normal selves.

Carl comes home from school with a two-foot Douglas fir seedling in a plastic bag with roots wrapped in a wet paper towel. It was an award, he tells me, for getting his desk organized quickly. He has a snack and the two of us put on rain gear and go outside to plant it.

"You know this guy in school today," Carl says, "he said a new tree should be planted for every tree that is cut down."

"At least," I say, thinking of all the trees that have been cut without any replanting. "Do you know," I ask him, "that this whole area, the roads, the fields, the farms, our vineyard, all used to be part of one big forest just a little over a hundred years ago?" He nods. The two of us walk down into our little grove behind the house and look for a good spot. Except for a few broadleaf maples, hawthorns, and wild cherry trees, there is only brush and an impenetrable jungle of blackberry vines. Slowly rotting Douglas fir stumps from the original logging dot the slope. Next to an old stump we find a natural clearing and I suggest that we plant it there. "This must be a good place for a tree," Carl says, "since this one grew so big here before." We clear the ground and cut a few branches above to eliminate shading in the spring. The hole is quickly dug and the tree planted. As we leave Carl asks, "Do you know what the most important thing is when you plant a tree?"

Suspecting nothing, I reply, "A shovel?"

"No," he says, "a tree."

When Cynthia walks down to get the morning paper, she finds our mailbox whacked off its wooden post. It looks as if it has been struck with a board; this is the third time it has happened. The box is down in the ditch, the lid twisted and torn, just barely attached to one of the hinges. This time it is beyond repair.

As a replacement, I get one made of quarter-inch steel, weighing well over twenty pounds, and haul it down to a local metal shop to have it welded onto a six-inch steel pipe. The welder tells me about friends who had their mailbox smashed and did what we are doing, only to find the whole thing pulled out of the ground. So he has welded spikes on to the pipe—free of charge—to better anchor it in the ground. I doubt it will prevent someone serious enough to pull it out, but there is enough iron there to require two adults just to load into the pickup.

One day when a friend is helping me move a large, cast iron stand for the wine press, a neighbor I hardly know comes half walking, half running through the vineyard. She must have climbed the fence. Between breaths she explains that one of their horses has been kicked, and a front leg is broken. Now the vet has just arrived but needs help to hold the horse down. Her husband can't stand the sight of blood and is of no help. Could we please come and assist them?

We leave the stand in the middle of the driveway and hurry back to her place the same way she had come. Inside the barn, the horse's hind legs are tied. The vet is a fat, older man with a round, clean-shaven face, dressed in dirty overalls, straw hat, and a thick sweater that seems much too large. He does not look like a vet. His face is shiny with sweat. He turns to the woman and points to a little cylinder tied around his neck. "If I pass out," he wheezes, as if he too had been running, "put a nitro tablet under my tongue."

She nods, and he turns to us, explaining how to throw the horse right, and where and how to hold him once he is down. We manage to get the horse down and hold it, but the man has trouble getting the splint set. He keeps cursing under his breath. At least he does not pass out. I have never hugged a horse like this before. The body is warm and strong and I like the smell of horse and straw. I start thinking about what it would be like to farm with a horse instead of a tractor: Would the right kinds of implements be available for vineyard work? How much pasture would it need? And where would I put it in the wintertime— I'd have to build a barn for it. "I don't know," the vet pants, interrupting my horse dream, "it's a bad fracture and it might not heal up." The horse is released, and we all stand to the side. Nobody says anything, knowing the consequences.

Afterward, the woman asks us into the office where she takes out three shot glasses and a bottle of bourbon from a wooden cabinet on the wall. She fills our glasses, we nod to each other and to her and swig it down in silence. I have never been inside here before; I did not even know that this building was an office. There is no sign of the husband. On the walls there are about a dozen picture frames with European paper money from the forties. I recognize marks, francs, lire, pounds, pesetas, kroner. "Yes," the woman says, noticing my interest in them, "my husband brought those back after the war. Souvenirs from all the countries he was stationed in. We got married as soon as he came home."

Cynthia notices that one of the hens has turned into an egg eater. Perhaps it has to do with the fact that some Orpingtons do not seem to like nesting boxes, but prefer to lay their eggs directly on the ground in a corner somewhere, leaving them exposed to pecking. If we do not remove the eggs immediately, the hens might break them. As we watch, we observe that this hen has learned to recognize that an egg is about to be laid, and patiently waits next to the hen laying it. As soon as the egg is on the ground, she immediately starts pecking at it until it breaks, triggering a frenzy of fighting hens trying to get pieces of the shell. All books agree that there is no cure for egg eaters, so it is time to have chicken for dinner again.

There is a large halo around the full moon when I go to bed. It has a dirty, brownish hue and I wonder what strange particles are in the atmosphere tonight. There is a breeze from the southwest sharpening itself on the corner of the house. With every gust of wind, the downspout hums a low vibration. Coyotes are yipping and howling across the road to the east, and the neighbor's dog barks and barks.

In the morning I run into our neighbor. He is returning from a walk carrying a bag with empty beer and pop cans.

"Did you see the halo around the moon last night?" I ask.

"No, but those damned coyotes kept me awake for hours. They know I've got sheep in that barn.

"Don't you think there are more coyotes around here now?" he continues before I even have a chance to open my mouth. "When people had cattle they kept their fences mended and they trapped the coyotes. Now, the fences are just falling down everywhere and the coyotes can run as they please."

"Well," I say, not really knowing how it was before, "maybe there are more of them now, but some people must still trap them. I saw an old coyote carcass just up the road a couple of days ago."

"That was no coyote," he says, "that was someone's old German shepherd." Then he waves and continues down the hill, leaving me worried about the quality of our fence line.

One afternoon I realize that our cat Vidar is gone. No one has seen him for several days and there are no clues as to his whereabouts. We all look for him in his favorite spots, but they are empty. We call his name, but there is no answer. We scatter and comb our entire eight acres without finding anything. We walk along the road checking the ditches to see if he has been run over by a car, but there are only the usual fast-food containers and cheap, empty liquor bottles. We search his favorite wood to the west, but understand that we have never known his full range.

Vidar is an old Norse name with roots both in the words "woods" and "widely traveled." It was a fitting name, because Vidar always liked to slip into the forest and often used to nap along its edge in the shade under the young chestnut tree, or in a sunny spot by the wine cellar. He liked to roam, and being a good hunter, brought back the things he caught: gophers, moles, voles, shrews, mice, goldfinches, housefinches, robins, swallows, Oregon juncos, a young owl once, a juvenile hare too, even snakes.

We wonder what got him. A car still seems most likely; perhaps he was hit and managed to crawl off the road to try to hide and heal. Or was he shot by a neighbor? He would not be the first. Maybe he was caught by a bobcat or some other predator? The old man down by the creek likes to mention what he claims is a photograph of a cougar paw print in the mud, and like the old man, every hunter knows the give-and-take in the food chain.

Digging out the last pinot noir vines, I think about the decade of accumulated labor leading up to this. When the decision was made to eliminate the last of the plants, I felt upset. It was in August, when the grapes again showed signs of mildew in spite of spraying. I remember thinking about all the energy we had expended on them. How we had tilled the old pasture to prepare it for the vines; how we had planted and weeded, mulched and watered, staked them with bamboo, and then trained them, endlessly fighting the weeds around the trunks. The first years we had even hoed around each vine by hand. We had replaced the failures, set posts, and strung wires, and each spring and summer we had sprayed them with sulphur. And still we had never harvested any grapes.

As I work, I can see the red flames of a huge fire in the forest behind the Christmas trees across the road. It is a tree grower from Molalla, working with his son-in-law, burning windfall and slash generated from thinning his trees. The billowing white smoke quickly joins the low clouds. The man planted that forest when he was middle-aged, and he may never see the trees cut. He planted it in a pasture that the Kamraths had labored hard to clear from the original forest fifty years earlier.

Perhaps the pinot noir has not come to nothing. Perhaps these ten years are simply the price of the insight they finally gave me. Perhaps I was not paying attention to what the vines were telling me, or even if I heard something, maybe I did not understand what it meant. Without knowing, how can one even ask the right questions?

At dawn a white hush holds the landscape. It is as if I woke up in a new place today; the land has a new emphasis. There are still a few flurries in the air. The chickens look surprised and hesitantly peck at the snowflakes whirling through the poultry netting. The sky is low like a giant exhalation out of some enormous arctic lung. The colors have all changed. What I thought was white turns out to be grey, and in the presence of snow I notice how drab and dirty the world really is.

The powdery day turns into a crisp evening with a fairly good cloud cover, but with some scattered stars shining through. The darkness is not so dark when it has snowed. After closing the gate, I decide to walk through the vineyard just to hear the snow crunch under my feet again. The vines are dormant and may be pruned again; the vineyard cycle has completed another revolution. The trunks and canes are silhouetted and distinct like calligraphy, each black vine delicately outlined with a layer of white on top. Hundreds and hundreds of them in the rows, each like lines of characters in a new alphabet I am still struggling to learn.

Winter solstice comes and the planet tilts in a magical moment of repose, like that ancient Greek statue of the discus thrower, where motion is caught in the perfectly still instant of turning. It is not easy to find that balance, to know where the center is located. What I was, I no longer am, and where I came from, no longer exists. The old match is gone. It is clear that what brought me here is no longer the same as that which holds me. This is the turning point; now is always the turning point. Soon the discus will spin forward into the future and everything will start over, but changed. What I am now I will no longer be. The future is waiting within everything. Hidden inside the sleeping canes, another vintage is waiting for the weather to tell it what to become. Nils and Carl are listening, trying to hear the future. The seasons will turn. There will be new births and new deaths. New perspectives will rise up; other possibilities will present themselves.

I keep a small bonfire going in the early evening as a few friends arrive to celebrate the year we have had, and to talk about the times to come. Standing around the flames, we drink sweet, mulled wine with almonds and raisins. The boys char marshmallows, their smell mingling with the smoke. One winter it snowed on the solstice and there was a peaceful hush around the fire, deeper and more hermetic than the silence of cathedrals.

Later, when the guests are gone and the house is quiet, I wait for the moment of turning in an easy chair by the wood stove. The exact time of the solstice passes, and I know that the light will slowly gain momentum again, starting to push back the darkness.

In the night, the east wind begins to howl out of the Columbia River Gorge again, a dry, cold, persistent hum. I wake up, and before I fall asleep again, I listen to this river of air. Whatever it can grab it throws around, making noises in the dark. It pushes over a pile of plywood scraps. It picks up empty aluminum cans and milk jugs from our recycling box, and kicks them across the gravel parking lot. Every now and then it rumbles in the fiberglass roofing over the chicken coop, and with a rattling sound it tugs and tugs at the dry, leathery leaves on the young oak tree, working hard to tear them off.

In the morning, the wind is still blowing and I find that it has piled tumbling weed seeds from far away along the bottom of all the north-south fence lines. It has delivered a Christmas card addressed to our neighbors to the door of the wine cellar. I quickly return inside and put on some tea water. From the window I watch the Oregon juncos search for food. They are small and tough; they have been here a long time and they know what it takes—staying close to the ground and facing the wind.

Lars Nordström was born in 1954 in Stockholm, Sweden where he lived until he was twenty. He was educated at the University of Stockholm, Portland State University, and Uppsala University, where he received his Ph.D. in American literature. He has been the recipient of several Fulbright grants, a Scandinavian Foundation grant for academic research in the United States, and several Swedish Institute grants and awards, as well as a Rockefeller Foundation Bellagio Center fellowhip. Nordström has published prose, poetry, translations, interviews, articles, and scholarly materials in Sweden, Finland, Canada, and the United States.

M. J. Pfanschmidt has been exhibiting her work since 1982. An avid printmaker as well as a promoter of prints and printmaking, she has served on numerous boards and committees of print organizations. She also works in various drawing media and egg tempera, and creates small-scale sculpture. She teaches drawing and printmaking at Marylhurst College in Lake Oswego, Oregon, and in 1998 she will receive her MFA from Vermont College.